ADVANCES IN INTELLIGENT MATERIAL SYSTEMS AND STRUCTURES

Volume 2

Advances in Electrorheological Fluids

EDITED BY MELVYN A. KOHUDIC

CRC Press
Taylor & Francis Group
Boca Raton London New York

CRC Press is an imprint of the
Taylor & Francis Group, an **informa** business

CRC Press
Taylor & Francis Group
6000 Broken Sound Parkway NW, Suite 300
Boca Raton, FL 33487-2742

© 1994 by Taylor & Francis Group, LLC
CRC Press is an imprint of Taylor & Francis Group, an Informa business

No claim to original U.S. Government works

Visit the Taylor & Francis Web site at
http://www.taylorandfrancis.com

and the CRC Press Web site at
http://www.crcpress.com

TABLE OF CONTENTS

PUBLISHER'S NOTE: All of the material in this publication has been reprinted from copyrighted articles in the *Journal of Intelligent Material Systems and Structures*, 1992, 1993 and 1994.

Response of Electrorheological Fluid-Filled Laminate Composites to Forced Vibration

Y. CHOI, A. F. SPRECHER AND H. CONRAD
Materials Science and Engineering Department
North Carolina State University
Raleigh, NC 27695-7907

ABSTRACT: Laminate composite beams consisting of metal or polystyrene outer strips and an electrorheological (ER) fluid (corn starch in corn oil) filler were constructed and tested using the standard Oberst test for viscoelastic materials. Application of an electric field of 2 kV/mm increased both the frequency of the various resonance modes and the loss factor associated with each. The apparent elastic modulus and loss factor of the composite beam and those of the ER fluid alone were calculated employing standard ASTM equations. Both modulus and loss factors of the composite beams decreased with resonance mode or frequency. The calculated shear modulus of the ER fluid increased with frequency and the loss factor decreased, their magnitudes being in reasonable accord with those measured directly on the fluid employing a rheometer. However, the calculated magnitude of the increase (GPa) of the apparent modulus of the metal/ER fluid composite beams with electric field and the fact that it decreases with frequency suggest that the standard ASTM equations do not adequately describe the vibration characteristics of the ER fluid laminate composites. A relatively simple on-off control system is illustrated as a potential application of ER fluid composites for vibration control in structural materials.

INTRODUCTION

LAMINATE composites consisting of structural materials separated by an electrorheological (ER) fluid have been previously prepared and exploratory studies on their mechanical properties reported (Coulter and Duclos, 1990; Choi, Sprecher, and Conrad, 1990). The rheological properties of the ER fluid can be adjusted through the application of an external electric field. These rheological changes are fast ($\approx 10^{-3}$ s) and reversible, thereby making the fluids suitable for real-time control of vibration. The controllability these fluids bring to structures and mechanical devices makes them suitable for smart materials systems or adaptive structures which can be adjusted *in situ* to accommodate a multiple range of excitation frequencies.

This paper presents the results of an investigation into the response of ER fluid-filled laminate beams in forced oscillation. The standard Oberst test (ASTM, 1983) was employed to explore the viscoelastic behavior of the laminates as a function of driving frequency, applied electric field and structural material in the outer layers of the composite. Preliminary results were given in the paper by Sprecher, Choi, and Conrad (1991).

EXPERIMENTAL

The ER fluid-filled composite was prepared in a manner similar to that described in Choi, Sprecher, and Conrad (1990). Laminate beams were produced by sandwiching an ER fluid between outer layers of a structural material (polystyrene, aluminum alloy and 70/30 brass) and sealing

with silicone rubber. Dimensions and some properties pertaining to the constituents of the beam are given in Table 1. The rheological properties of the ER fluid in the post-yield region are given in Choi, Sprecher, and Conrad (1990).

A standard Oberst test and equations (ASTM, 1983) were employed to generate and analyze the data. A schematic of the test apparatus is given in Figure 1. The beam was clamped at one end and driven by a non-contacting electromagnetic actuator (sinusoidal profile, 5 to 250 Hz). Beam response was monitored through a non-contacting proximity probe positioned at one quarter of the beam's length from the clamped end. All tests were performed at room temperature.

RESULTS

The transmissibility response of the composite beam was determined from oscilloscope traces of driving force and displacement; see, for example, Figure 2 for the aluminum-ER fluid composite beam. As can be seen, the resonance frequencies are shifted to higher values with application of the electric field as a result of the change in the rheology of the ER fluid. Also, there occurred a slight widening of the -3 dB bandwidth, indicating an increase in the damping characteristics of the beam with electric field. The effect of the electric field on the resonance frequency for each resonant mode and the modal loss factor $\eta_s = \Delta f_n / f_n$ (where f_n is the resonance frequency of the nth mode and Δf_n is the bandwidth at -3 dB) are presented in Figure 3 for the aluminum-ER composite beam. Table 2 summarizes the results for the various beams.

1

Table 1. Dimensions and some properties pertaining to the composite beams.

Material	E (GPa)	ϱ (g/cm³)	H (cm)	L (cm)	b (cm)	H_1 (cm)
Al alloy (98.8% Al—1.2% Mn)	70	2.70	0.03	28.8	3.2	
Brass (70% Cu—30% Zn)	110	8.50	0.04	28.8	3.2	
Polystyrene (with Al tape electrode)	3	1.06	0.06	28.8	3.2	
Silicone rubber	0.003	1.50				0.18*
ER fluid: 34 wt% corn starch		1.06				0.18**
8 wt% H₂O in corn oil						

*10 vol% of combined ER fluid + silicone rubber.
**90 vol% of combined ER fluid + silicone rubber.
H = thickness of outer sheets; L = length of beam; b = beam width; H_1 = gap spacing.

The *apparent* Young's storage modulus E_s' and loss factor η_s of each composite beam was determined from the resonant frequency and band half-width through the following relations (ASTM, 1983)

$$E_s' = [12\varrho L^4 f_n^2]/[H^2 C_n^2] \qquad (1)$$

where ϱ is the average density of the Oberst bar, L the length of the bar, f_n the resonance frequency for mode n, H the bar thickness and C_n the coefficient of mode n due to the boundary condition of the vibrating beam. In addition to the *apparent* storage modulus, the loss factor $\eta_s (= E_s''/E_s'$, where E_s'', the loss modulus, is the imaginary part of the complex modulus) was determined from the half-power bandwidth.

The *apparent* storage modulus E_s' versus frequency is presented in Figure 4(a) for the aluminum-ER fluid composite beam. The polystyrene and brass composite beams exhibited similar behavior. The results for all of the ER fluid composites investigated are summarized in Table 3. Evident is that the *apparent* storage modulus increased by 25%–100% upon application of the electric field. Moreover, the modulus both with and without the electric field decreased with resonant frequency.

The effect of electric field on the loss factor η_s is also included in Tables 2 and 3. The loss factor η_s reflects the damping or energy dissipating property of the beam. The electric field produced increases in the loss factor of 42% to 133%. Furthermore, η_s decreased with increasing frequency as did the apparent modulus.

Also of interest are the modulus components of the ER fluid alone under the present test conditions. The ER fluid shear modulus, G_2' and its loss factor η_2 were determined using the following relationships for a sandwich beam (Ross, Ungar, and Kerwin, 1959; ASTM, 1983; Nashif, Jones, and Henderson. 1985).

$$G_2' = [(A-B) - 2(A-B)^2 - 2(A\eta_s)^2] \cdot [2\pi C_n E_{os}' H H_2/L^2]/$$

$$[(1 - 2A + 2B)^2 + 4(A\eta_s)^2] \qquad (2)$$

and

$$\eta_2 = (A\eta_s)/[A-B-2(A-B)^2 - 2(A\eta_s)^2] \qquad (3)$$

where the subscripts s and os refer to the sandwich (com-

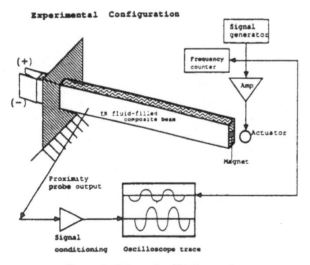

Figure 1. Schematic of test apparatus.

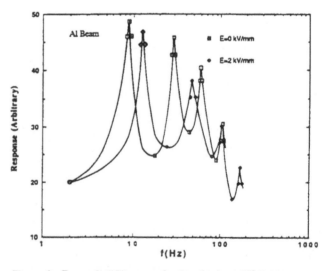

Figure 2. Transmissibility curve for the aluminum/ER fluid composite beam.

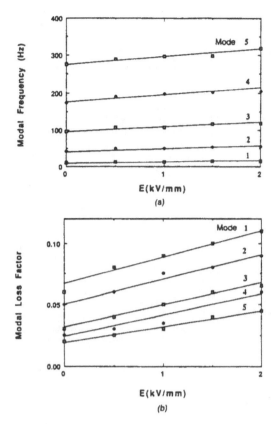

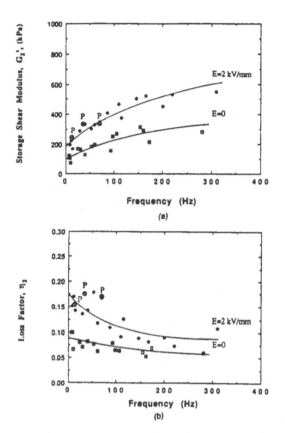

Figure 3. *Effect of electric field at increasing resonance modes on: (a) modal frequency and (b) modal loss factor.*

Figure 5. *Effect of electric field on the mechanical properties of the ER fluid in the aluminum ER composite beam: (a) apparent storage shear modulus G_2' and (b) loss factor η_2. Data points labeled P were obtained on the fluid alone using a rheometer and were taken from plots similar to Figure 6.*

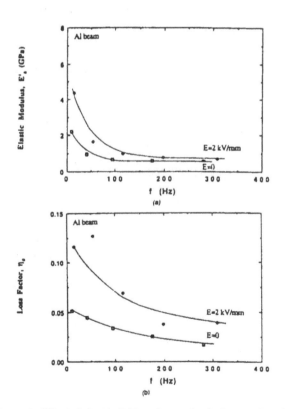

Figure 4. *Effect of electric field on the mechanical properties of the aluminum/ER fluid-filled composite beam: (a) apparent elastic modulus E_s' and (b) loss factor η_s.*

posite) beam and outer strip material respectively, $A = (f_s/f_{os})^2(2 + DT)(B/2)$, $B = 1/[6(1 + T)^2]$, $T = H_2/H$, $D = \varrho_2/\varrho$. ϱ_2 and ϱ are the density of the damping material (ER fluid and silicone rubber) and outer strip materials, respectively, and $\eta_s = \Delta f_s/f_s$.

The variation of G_2' and η_2 with frequency of vibration f for the Al- and brass-ER fluid composite beams is given in Figure 5 and for all sheet materials in Table 4. Evident is that G_2' increases with f and η_2 decreases, as expected for the ER fluid. The difference in G_2' (and η_2) between the value with the field and that without should represent that of the ER fluid per se. As a check on this, G_2' and η_2 of the ER fluid were determined directly using a Physica universal rheometer. Figure 6 gives log-log plots of G_2' and η_2 obtained as a function of the oscillating shear strain γ at a frequency of 30 Hz. To be noted, G_2' increases, whereas η_2 decreases with γ. The shear strain in the composite beams during the forced oscillation was calculated to be $\sim 3 \times 10^{-4}$. Therefore, the plots in Figure 6 were extrapolated to this strain and the results are given in Figure 5 by the data points lettered P. Reasonably good agreement exists between the extrapolated values of G_2' and η_2 obtained using the Physica rheometer and those calculated from the forced vibration of the composite beam using the standard ASTM equations.

Table 2. *Effect of electric field on the forced vibration characteristics of the ER fluid-filled composite beams.*

| Outer Sheet Material | Resonance Mode Frequency, f_n (Hz) | | | | | | | | | |
| | Mode I | | Mode II | | Mode III | | Mode IV | | Mode V | |
	$E = 0$	$E = 2$	$E = 0$	$E = 2$	$E = 0$	$E = 2$	$E = 0$	$E = 2$	$E = 0$	$E = 2$
Al alloy	8.5	11.4	29.5	39.9	60.4	86.4	105.8	146.2	161.7	220.0
Brass	6.7	8.1	24.0	28.7	52.9	60.6	97.3	109.5	154.5	167.0
Polystyrene	6.1	7.6	23.7	29.7	48.5	62.7	82.2	103.6	N/A	N/A

| Structural Material | Modal Loss Factor, η_s | | | | | | | | | |
| | Mode I | | Mode II | | Mode III | | Mode IV | | Mode V | |
	$E = 0$	$E = 2$	$E = 0$	$E = 2$	$E = 0$	$E = 2$	$E = 0$	$E = 2$	$E = 0$	$E = 2$
Al alloy	0.08	0.12	0.06	0.12	0.04	0.09	0.04	0.07	0.03	0.05
Brass	0.09	0.12	0.05	0.10	0.03	0.06	0.02	0.04	0.02	0.03
Polystyrene	0.11	0.13	0.12	0.15	0.09	0.14	0.07	0.10	N/A	N/A

Table 3. *Effect of electric field on the viscoelastic properties of the ER fluid-filled composite beams.*

| Outer Sheet Material | $E = 0$ | | | | $E = 2$ kV/mm | | | |
| | E_s' (GPa) | | η_s | | E_s' (GPa) | | η_s | |
	Mode I	Mode V	Mode I	Mode V	Mode I	Mode V	Mode I	Mode V
Al alloy	3.5	0.5	0.08	0.03	6.5	1.0	0.13	0.07
Brass	5.0	0.8	0.08	0.02	7.5	1.0	0.12	0.03
Polystyrene	1.5	0.3	0.12	0.08	2.5	0.5	0.17	0.12

Table 4. *Effect of electric field on the viscoelastic properties of the ER fluid derived from the vibration response of the composite beams.*

| Outer Sheet Material | $E = 0$ | | | | $E = 2$ kV/mm | | | |
| | G_s' (GPa) | | η_2 | | G_s' (GPa) | | η_2 | |
	Mode I	Mode V	Mode I	Mode V	Mode I	Mode V	Mode I	Mode V
Al alloy	0.10	0.30	0.10	0.05	0.20	0.60	0.15	0.10
Brass	0.10	0.30	0.10	0.05	0.20	0.55	0.15	0.10
Polystyrene	0.05	0.12	0.17	0.13	0.09	0.25	0.23	0.17

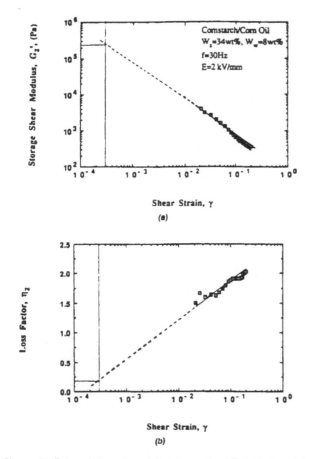

Figure 6. *Extrapolation of moduli data on the ER fluid alone (obtained using a Physica rheometer) to the shear strain experienced by the composite beams.*

DISCUSSION

The present results reveal that the forced vibration characteristics of a laminate composite beam consisting of an ER fluid sandwiched between two structural sheet materials can be significantly altered upon application of an electric field to the fluid. An electric field of 2 kV/mm increased the frequency at which each of the resonant modes occurred and decreased their amplitude or transmissibility. Moreover, it produced a widening in the band half-width.

Employing the standard ASTM equations for a vibratory Oberst bar the change in resonant frequency produced by the electric field represented a 25% to 100% change in the *apparent* storage elastic modulus of the composite beam and the increase in half-width represented a 42 to 133% change in the loss factor. Unexpected, however, was that the derived moduli of the composite beam decreased with increase in resonant frequency, since the modulus of the outer sheet materials was independent of frequency (Sprecher, Choi and Conrad, 1991) and the modulus of the ER fluid increased with frequency (Figure 5). This anomalous behavior probably results from the fact that the deformation of the composite beam is not completely in accord with the assumptions in the ASTM equations. The equations used in the standard Oberst test assume a homogeneously deforming, viscoelastic material and relate the resonant frequency to the beam geometry and test conditions (Ross et al., 1959; ASTM, 1983). In the present experiments the viscoelastic intermediate layer consists of the ER fluid with an anisotropic structure and the silicone rubber used to seal the system. If any nonuniform deformation occurs in the beam, the Equation (1) used to determine the modulus of the composite beam from the transmissibility curves would not be entirely correct. In view of this, it is somewhat surprising that Equations (2) and (3) yielded values for G_2' and η_2 of the ER fluid which were in reasonable agreement with those determined from direct measurements on the fluid with the Physica rheometer.

Also surprising was the fact that an increase in the shear modulus of the ER fluid filler of only about 10^5 Pa obtained upon application of the electric field (Figure 6) could produce changes in the vibration characteristics of the metal-ER fluid composite beams equivalent to an increase in their elastic modulus of the order of GPa (Figure 4). Again, this indicates the need to review the applicability of the Oberst equations to these composite beams.

Considering the transmissibility curves of Figure 2, the properties of an ER fluid-filled composite could be used to advantage with a simple on-off control system shown in Figure 7, similar to that suggested by Coulter and Duclos (1990). This scheme could be employed to shift the beam's resonance and thereby reduce the total system response over a broad frequency range.

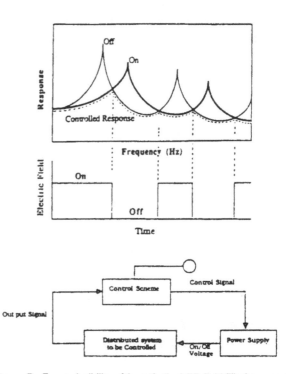

Figure 7. *Transmissibility of hypothetical ER fluid-filled composite material using an on/off control strategy. Light solid line is for the composite with electric field off; heavy solid line is with field on; dashed line is using on/off control.*

SUMMARY AND CONCLUSIONS

1. Composite beams consisting of a laminate of metal or polystyrene outer strips and an ER fluid of corn starch in corn oil were constructed and tested in forced vibration using the standard Oberst test for viscoelastic materials. It was found that the resonant frequency of all modes and the associated bandwidth increased in a linear fashion with electric field.

2. Analysis of the data using the standard Oberst equations gave increases in the *apparent* elastic modulus E_i' of the composite beams of 25 to 100% and in the loss factor η_a of 42 to 133% upon application of an electric field of 2 kV/mm, the effect decreasing with increase in resonance mode or frequency. In the case of the metal laminates, the magnitude of the increase in modulus was of the order of GPa.

3. The shear modulus G_2' and loss factor η_2 of the ER fluid filler were also determined using the standard Oberst equations and were found to be in reasonable accord with those measured directly on the fluid alone employing an oscillating rheometer. G_2' was of the order of 10^5 Pa and increased with frequency, whereas η_2 was of the order of 0.05 and decreased with frequency.

4. The following two results suggest that the standard Oberst equations do not adequately describe the vibration characteristics of the ER fluid-filled laminate composites:

   (a) Changes of the order of 10^5 Pa in the modulus of the ER fluid filler produced by the electric field resulted in changes of the order of 10^9 Pa in the modulus of the composite beam.

   (b) The elastic modulus of the composite beams decreased with increase in frequency, whereas the modulus of the outer layers were independent of frequency and that of the ER fluid filler increased with frequency.

5. A simple on-off control system is described which could be used to control the vibration characteristics of ER fluid composite beams in structural applications.

ACKNOWLEDGEMENTS

The authors gratefully acknowledge support of this work by the National Science Foundation under NSF award #CBT-8714515, the Ford Motor Company and the NCSU University/Industry Consortium on ER fluids.

REFERENCES

1. ASTM. 1983. "Standard Method for Measuring Vibration-Damping Properties of Materials", ASTM Standard E-756-83, *The American Society for Testing Materials*, Philadelphia, PA.

2. Choi, Y., A. F. Sprecher and H. Conrad. 1990. "Vibration Characteristics of a Composite Beam Containing an Electrorheological Fluid", *J. Intel. Mat. Syst. and Struc.*, 1(1):91–104.

3. Coulter, J. P. and T. G. Duclos. 1990. "Applications of Electrorheological Materials in Vibration Control", J. D. Carlson, A. F. Sprecher and H. Conrad, eds., in *Proc. 2nd Int. Conf. Electrorheological Fluids*, Lancaster, PA: Technomic Publishing Co., pp. 300–325.

4. Nashif, A. D., D. I. G. Jones and J. P. Henderson. 1985. *Vibration Damping*, New York: John Wiley and Sons.

5. Ross, D., E. E. Ungar and E. M. Kerwin, Jr. 1959. "Damping of Plate Flexural Vibrations by Means of Viscoelastic Lamina", J. E. Ruzicka, ed., *Structural Damping*, Sec. 3, New York: The American Society of Mechanical Engineers.

6. Sprecher, A. F., Y. Choi and H. Conrad. 1991. "Mechanical Behavior of ER Fluid-Filled Composites in Forced Oscillation", B. K. Wada, J. L. Fanson and K. Miura, eds., in *First Joint U.S.-Japan Conference on Adaptive Structures*, Technomic, Lancaster, PA, pp. 560–579.

Comparative Methods for the Derivation of In Flow Electrical Characteristics of Electro-Rheological Fluids

A. Hosseini-Sianaki, R. Firoozian, D. J. Peel and W. A. Bullough

Department of Mechanical and Process Engineering
University of Sheffield
P.O. Box 600
Sheffield, SI 4DU, U.K.

ABSTRACT: Results are presented from two identification exercises which have been used to determine the resistance and capacitance values of a typical electro-rheological fluid. Simulated data, in which the parameter values are known a-priori, are employed to compare the two methods with regard to both accuracy and computational efficiency. The experimental data analysed are from a parallel plate flow mode valve restrictor. Both approaches produce similar results over the operating range of moderate to high excitation frequency, but yield substantially different results at low frequencies. Possible reasons for these inconsistencies and methods for resolving the problems are discussed. Amongst these problems noise on the processed signal is a major factor.

INTRODUCTION

LITTLE quantitative modelling of the all important time response domain of the electro-rheological fluid or Winslow Effect mechanism has so far been undertaken. Yet, if the effect is to be optimised in one of its most promising aspects, fast latches of one kind or another etc. (Bullough et al., 1991), not only must the controllable stress level be increased but it should be achieved quickly after the switch on of the excitation. These conditions can be characterised by two major quasi-independent time domain parameters, the electrical and mechanical/ acceleration time constants respectively. The latter is determined by the magnitude of the controllable yield stress available insomuch as it fixes the size and hence the inertia of the driven element for a given traction effort. The former has only been systematically investigated in the flow mode where it is seemingly the result of a lead phenomenon. Future progress in the applications and developments of fluids in the time domain requires a knowledge of all the underlying mechanisms (see Bullough, 1989 and Peel, 1989).

One way of achieving this is via fundamental studies into the properties of molecular electronic materials and the correlation of rheological structural moduli with mechanical and electronic/material characteristics. Such work is intensively multi-disciplinary, especially with respect to the time domain.

Another approach, to which the present paper is relevant, involves the identification of the physical mechanisms from high quality, fast and simultaneous recordings of the many variables involved. This is rather like a dimension analysis but, with numbers. The aim is to detect what combination of events could be responsible for the shape of the electro-rheological time response—by quantitative assessment (magnitude and sequence). However, like dimensional analysis, more than one result is possible—some are not real or in a useful format. All are highly non-linear in fashion.

In an attempt to break down the impasse caused by (so far) ineffective real flow/particle movement visualisation (the attendant high voltages and difficulties in achieving a satisfactory scaling of particle kinematics and dynamics causes problems) and with parameter diffusion, the initial part of such a step by step exercise has been restricted by the adoption of a constant flow and temperature regime. Also, in the first instance it deals with electrical performance alone. This clarifies the objective—to test the validity and accuracy of the approach, viz.: for the response of current to sine and step excitation in both the milli second and long duration time regimes.

Eventually the aim will be, given encouragement in this initial exercise, to converge with the results of other workers. This will bring together industrial range ER controller findings with the dynamics of materials and structure often found on more micro and molecular scale and at low shear rates respectively. A spin-off to be gained from the planning exercise associated with this work was to specify what data would be needed from a new shear mode apparatus and how it should be achieved.

The first step along this path was to construct a resistance/capacitance model of the electrical response from DC biased inputs. The values of R and C were identified over a range of small sine wave frequencies and (set) bias voltages and flow rates. Subsequently these were consolidated in a search for inductive effects (Firoozian et al., 1990) where they were used to predict the experimental response to a step input.

Given a secure electrical model, the only remaining model to be identified is the relationship between the applied current and shear stress. In the present case flow mode results were used in order to keep the temperature within

7

close limits. The rheological/semi-conductor amalgam like its solid phase counterpart doubles its conductance for every 5 to 10°C temperature change. Pressure was thus the outstanding uncontrolled variable. For the present it is assumed to be independent of the voltage/current relationship and so plays no part at this stage of affairs. There is some suggestion that this assumption may not be totally valid. The presence of sympathetic noise on the current and pressure signals suggests some pressure/impedance interaction (Bullough, 1989).

Thus far the results have been encouraging. Quality measures adopted for the control of transducer data appear to have been successful. Nevertheless, it is considered appropriate at the outset of a large data processing exercise into the identification of unknown physical phenomena to check that the first result is valid beyond all reasonable doubt. It is this part of the work that is described in the present paper. It includes an assessment of the effect of anti-aliasing and noise reduction procedures used to prevent corruption of the multi-channel readings.

This objective is attained by introducing an alternative identification technique [to the one employed in (Firoozian et al., 1989)], namely, the state variable filter (SVF) and performing some mathematical and experimental simulation to compare the two methods. Then, the experimental data from an ER valve rig is analysed and conclusions are drawn with regard to the accuracy and computational efficiency of the two techniques. In particular, special attention is given to the valve response at low excitation frequencies.

THE SYSTEM MODEL

A schematic diagram of the equivalent assumed linearised electrical circuit of the ER fluid is shown in Figure 1. The differential equation characterizing the relationship between the supplied voltage v and the current i may be written as

$$RC\frac{dv}{dt} + v = Ri \tag{1}$$

It is convenient to non-dimensionalise Equation (1) by introducing the variables $V = v/V_m$, $I = i/I_m$ and $\tau = \omega_e t$, where V_m and I_m are reference voltage and current respec-

tively, and ω_e is a reference frequency, e.g., excitation frequency. On substituting these into Equation (1) it can be written, viz.:

$$\frac{dV}{d\tau} + aV = bI \tag{2}$$

where

$$a = \frac{1}{RC\omega_e} \tag{3}$$

$$b = \frac{I_m}{C\omega_e V_m} \tag{4}$$

THE IDENTIFICATION PROCESS

The simplest method to evaluate the values of R and C in Equation (1) is to directly measure the amplitude ratio and phase angle from the voltage and current signals. For harmonic excitation, the governing differential Equation (1) takes the form of

$$\frac{i}{v} = \frac{RC(j\omega) + 1}{R} \tag{5}$$

with amplitude ratio

$$M = \frac{\sqrt{[(RC\omega)^2 + 1]}}{R} \tag{6}$$

and phase angle

$$\phi = \tan^{-1}(RC\omega) \tag{7}$$

The capacitance and resistance can be estimated by direct measurements of amplitude ratio and phase angle, i.e.

$$R = \frac{\sqrt{(1 + \tan^2 \phi)}}{M} \tag{8}$$

$$C = \left(\frac{\tan \phi}{\omega}\right)/R \tag{9}$$

The problem associated with this technique is that, for correct evaluation of R and C one requires very accurate measurements of current and voltage amplitudes and the phase angle between them. Given the fact that the experimental signals are noisy and that also the phase angle decreases by reducing the excitation frequency (see Figure 2), the accuracy in direct measurement is quite low. One approach in which the above problem is avoided is known as "parameter identification".

The non-dimensional Equation (2) describes the voltage-current relationship for a small perturbation of the voltage.

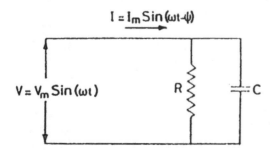

$I = I_m \sin(\omega t - \psi)$

$V = V_m \sin(\omega t)$ R C

Figure 1. Linearised equivalent electrical circuit for ER fluid.

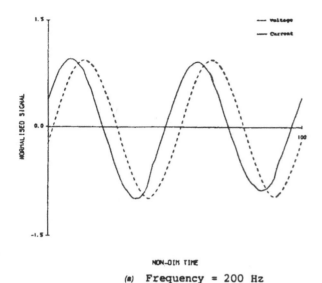

(a) Frequency = 200 Hz

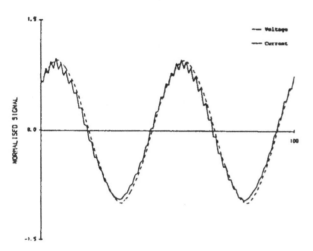

(b) Frequency = 10 Hz

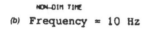

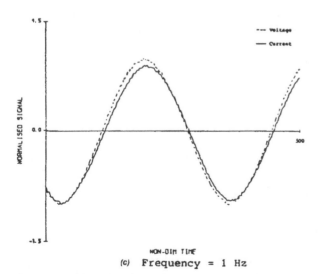

(c) Frequency = 1 Hz

Figure 2. *Current response to sinusoidal voltage excitation, valve data: (a) frequency = 200 Hz, (b) frequency = 10 Hz, (c) frequency = 1 Hz.*

By means of experimental trials it is possible to capture simultaneous records of the voltage and current at appropriate sampling intervals. Given these records the identification problem can be stated as follows: from the records $V(t_j)$, $I(t_j)$, where $j = 0, 1, \ldots, N$, how can one achieve estimates of the parameters a and b.

In a previous publication (Firoozian et al., 1989) a nonlinear identification algorithm, namely, invariant imbedding (II) (see Detchmandy and Sridhar, 1966) was employed and estimates were found for the resistance and capacitance of the ER fluid. Here, an alternative (linear-in-the-parameter) identification technique is introduced. It is the state variable filter (SVF) which, as will be shown later, is generally simpler and more accurate, than the II method and needs considerably less computational time.

Before proceeding any further with the SVF method it is appropriate to briefly describe the algorithm involved in the invariant imbedding technique.

The Invariant Imbedding Technique

To implement this identification technique the first step is to write Equation (2) in a state-space form. By considering the unknown parameters a and b as state variables, in addition to the system variables, one can rewrite Equation (2) as

$$\left. \begin{aligned} \dot{x}_1 &= x_2 I - x_3 x_1 \\ \dot{x}_2 &= 0 \\ \dot{x}_3 &= 0 \end{aligned} \right\} \qquad (10)$$

where $x_1 = V$, $x_2 = b$ and $x_3 = a$. Or in vector form:

$$\underline{\dot{x}} = \underline{f}(t,x) \qquad (11)$$

where

$$\underline{x} = [x_1\ x_2\ x_3]^T$$

and

$$\underline{f}(t,x) = \begin{bmatrix} x_2 I - x_3 x_1 \\ 0 \\ 0 \end{bmatrix}$$

Let the measurements be denoted by:

$$\underline{y}(t) = \underline{h}(t,x) + \text{measured noise} \qquad (12)$$

Equation (11) forms the basis for the implementation of the identification method of (Detchmandy and Sridhar, 1966) and may be summarised as:

$$\frac{d\hat{x}}{dt} = f(t,\hat{x}) + 2PHQ[y(t) - \hat{x}_1] \qquad (13)$$

and the matrix $\underline{P}$ is obtained as:

$$\frac{d\underline{P}}{dt} = \frac{\partial f(t,\hat{x})}{\partial \hat{x}} \cdot P + P \cdot \frac{\partial f^T(t,\hat{x})}{\partial \hat{x}} - 2PHQHP^T \quad (14)$$

where

$$\underline{P} = \begin{bmatrix} P_{11} & P_{12} & P_{13} \\ P_{21} & P_{22} & P_{23} \\ P_{31} & P_{32} & P_{33} \end{bmatrix}, HQH = \begin{bmatrix} 1 & 0 & 0 \\ 0 & 0 & 0 \\ 0 & 0 & 0 \end{bmatrix}$$

$\dfrac{\partial f(t,\hat{x})}{\partial \hat{x}}$ is the Jacobian of the vector $\underline{f}(t,\hat{x})$

x_1 and $\hat{x}_1$ are the measured and estimated voltage respectively.

An estimate of the state vector x denoted by $\hat{x}$ in Equation (13) can be generated as a function of time, by solving Equations (13) and (14) numerically, using the fourth order Runge-Kutta algorithm.

It is necessary to set initial values for x and the covariance matrix $\underline{P}$ to start the integration procedure. In practice one chooses $\underline{P} = \lambda J$ where J is the unit matrix and λ is the largest number which will not cause numerical instability. The choice of the large λ reflects the uncertainty of the parameter estimates. Having found steady values for x_2 ($=a$) and x_3 ($=b$) it is then easy to retrieve values for R and C from Equations (3) and (4).

The State Variable Filter Technique

It is believed that the pioneering work on the SVF approach was undertaken by Young (see Young, 1984) about three decades ago. However, its revival and application to engineering problems is mainly due to Gawthrop and his work on self-tuning control systems (see Gawthrop, 1984a, 1984b). In the past few years this method has successfully been applied to many fields of engineering including squeeze-film bearings (see Ellis et al., 1988 and Roberts et al., 1990) and ship roll motion (see Gawthrop et al., 1988).

By introducing the a-priori known constant c such that

$$a = c + f \quad (15)$$

Equation (2) may be represented as

$$\frac{dV}{d\tau} + cV = bI - fV \quad (16)$$

The above rearrangement has resulted in placing the unknown constants b and f on the right hand side of the equation. More importantly the voltage V now depends linearly on the parameters f and b. Constants b and f are found initially and using Equation (15), a is then recovered.

Consider the following set of linear auxiliary equations

$$\frac{dz_1}{d\tau} + cz_1 = -V \quad (17)$$

$$\frac{dz_2}{d\tau} + cz_2 = 0 \quad (18)$$

$$\frac{dz_3}{d\tau} + cz_3 = I \quad (19)$$

The process of (a) multiplying Equation (17) by f, (b) multiplying Equation (18) by an unknown constant d, (c) multiplying Equation (19) by b and adding (a) to (c) above together, gives

$$\frac{d}{d\tau} (fz_1 + dz_2 + bz_3) \quad (20)$$

$$+ c(fz_1 + dz_2 + bz_3) = bI - fV$$

Comparing Equation (16) with (20) gives

$$V = fz_1 + dz_2 + bz_3 \quad (21)$$

Using vector notation Equation (21) can be stated more concisely as

$$V(t) = \underline{y}^T(t)\underline{\theta} \quad (22)$$

where

$$\underline{y}(t) = [z_1 \ z_2 \ z_3]^T \quad (23)$$

and

$$\underline{\theta} = [f \ d \ b] \quad (24)$$

In Equation (21) the parameter d has been introduced to allow for the, in general, non-zero initial condition. In fact the initial conditions for the auxiliary differential Equations (17) to (19) are chosen such that

$$\left. \begin{matrix} z_1(0) = 0 \\ z_2(0) = 1 \\ z_3(0) = 0 \end{matrix} \right\} \quad (25)$$

then it is clear from Equation (21) that

$$d = V(0) \quad (26)$$

In practice, the experimental records of V and I taken at equispaced times $t_i = i\triangle t$ will contain measurement error. If the measured value of V is denoted by V_M then Equation (22) can be modified as

$$V_M(t_i) = \underline{y}^T(t_i)\underline{\theta} + \xi \qquad (27)$$

where ξ represents measurement noise.

The parameter vector $\underline{\theta}$ can be estimated by employing a least squares cost function which minimizes the sum of the squared differences between $V_M(t)$ values and the corresponding $\underline{y}^T(t)\underline{\theta}$ products for all the experimental samples. In other words,

$$J = \sum_{i=0}^{N} [V_M(t_i) - \underline{y}^T(t_i)\underline{\theta}]^2 \qquad (28)$$

Although standard algorithms are available for this purpose (Young, 1984), in the present paper, as in Roberts et al. (1990), a recursive algorithm is employed. This sequentially updates the estimate of the parameter vector $\underline{\theta}$, i.e., $\underline{\theta}(t_i)$, by marching in time steps through the voltage, $V_M(t_i)$ data records. However, prior to this the generation of the estimation filter state, $\underline{y}(t_i)$, records is required. This is achieved by solving the auxiliary Equations (17) to (19) using the Merson's Runge-Kutta numerical integration technique.

The recursive least-squares algorithm may be stated as follows: If $\hat{\underline{\theta}}$ is the least square estimate of $\underline{\theta}$ at time $t_N = N\triangle T$ and $\underline{y}_N \equiv \underline{y}(t_N)$, $V_N \equiv V_M(t_N)$, let

$$\underline{S}_N = \sum_{i=1}^{N} \underline{y}_i \underline{y}_i^T \qquad (29)$$

(a) $\quad \sigma_{N+1} = \underline{y}_{N+1}^T \underline{S}_N^{-1} \underline{y}_{N+1} \qquad (30)$

(b) $\quad \underline{K}_{N+1} = \dfrac{1}{(1 + \sigma_{N+1})} \underline{S}_N^{-1} \underline{y}_{N+1} \qquad (31)$

(c) $\quad \hat{e}_{N+1} = V_{N+1} - \underline{y}_{N+1}^T \hat{\underline{\theta}}_N \qquad (32)$

(d) $\quad \hat{\underline{\theta}}_{N+1} = \hat{\underline{\theta}}_N \underline{K}_{N+1} \hat{e}_{N+1} \qquad (33)$

(e) $\quad \underline{S}_{N+1}^{-1} = \underline{S}_N^{-1} - (1 + \sigma_{N+1}) \underline{K}_{N+1} \underline{K}_{N+1}^T \qquad (34)$

Steps (a) to (e) must be repeated by marching through time to achieve a final least-squares estimate for the vector $\underline{\theta}$.

To start the recursive estimation procedure the vector $\underline{\theta}$ is set to zero. The matrix, $\underline{S}_N^{-1}$ is analogous to the matrix $\underline{P}$ in the II technique and similarly is set to λJ (J is an identity matrix and λ is a large constant). However, as will be discussed later λ can be set to a much larger value than that for the II method, thus resulting in a much faster convergence of the parameters in the parameter vector $\underline{\theta}$.

THE SIMULATION TESTS

In an attempt to compare the performances of the two identification techniques with regard to accuracy and com-

putational efficiency, some simulations were carried out prior to analyses of actual experimental data from the ER valve. This section aims at reporting these trials.

As indicated in the preceding section the electrical model of the ER fluid comprises predominantly capacitive and resistive components in a parallel arrangement, any inductive component can be neglected. This latter point has been investigated in detail and findings are reported in Firoozian et al. (1990). These assumptions result in the model shown in Figure 1. To accomplish the objectives of this section two types of simulations were performed: Mathematical and Experimental.

Mathematical Simulation

To allow the proposed identification procedures to be comprehensively verified, they were first applied to digitally simulated data. In more detail, the differential equation describing the behaviour of the assumed mathematical model has the form of Equation (1). For systems with a-priori known values of coefficients R and C this equation was solved by an exact analytical method. The non-dimensionalisation explained earlier was then applied to the time series data records thus generated.

Since digital quantization error and electronic noise will invariably exist in the captured experimental data, the behaviour of the two identification algorithms in the presence of noise is a major consideration. In order to simulate the noise, Gaussian independent random numbers were added to simulated data using the pseudo-random number generator of a computer. The mean and standard deviation of the distribution were fixed to zero and unity respectively. Finally, since the amplitude of both current and voltage were normalised to unity the random numbers were multiplied by a fraction up to a value of ten percent before being added to the simulated data. It should be emphasised that these noise levels were considerably higher than those observed in the real experimental trials.

Extended runs with various values of R and C and also different excitation frequencies were carried out. Table 1 presents the results for a typical case when $R = 54$ kΩ and $C = 20$ nF. Two excitation frequencies were considered and in each case three levels of Gaussian noise were added to the digitally simulated data. It is evident from the table that both methods produce poor capacitance results at very low frequency and these are greatly affected by the noise level on the signals. However, it should be noted that SVF technique identifies both the capacitance and resistance very accurately in the absence of noise.

In contrast to low frequency responses, both approaches yield very accurate estimates of both capacitance and resistance at high excitation frequencies. It is also clear from Table 1 that the noise has very little effect on the identified parameters. It is worth pointing out the fact that the SVF results are in general more accurate than the II outcome in which the data has been reprocessed 100 times.

Table 1. Comparison between SVF and II, effect of noise on identified parameter values using mathematical simulated data. True values: C = 20 nF, R = 54 kΩ.

| Excitation Frequency | Noise Level | | SVF Method | | II Method | | | |
| | On Voltage | On Current | C | R | 1st Iteration | | 100th Iteration | |
					C	R	C	R
	0.0	0.0	19.96	53.98	231.5	54.08	116.6	53.7
2	0.01	0.05	100.67	54.01	233.1	54.12	122.04	53.82
	0.1	0.1	432.69	54.23	234.73	54.73	135.47	54.24
	0.0	0.0	20.02	52.79	27.29	56.01	21.21	54.02
60	0.01	0.05	20.14	52.7	27.45	55.96	21.23	54.05
	0.1	0.1	21.56	52.9	27.69	56.44	21.21	54.16

The reasons behind the difficulties in accurate estimation of the capacitance values at low excitation frequencies will be discussed in the context of the experimental data in the next section. A typical graphical example of the behaviour of the two identification algorithms is shown in Figure 3. The (a) sequence shows the SVF performance whereas the (b) sequence depicts the behaviour of the II method. In both cases the parameters a and b have fully converged and a

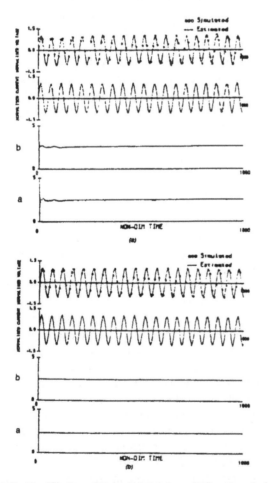

Figure 3. Identification of simulated data (a = ^{1}/RCω$_e$, b = I_m/Cω$_e$ V$_m$): (a) SVF method, response after one iteration, (b) II method, response after 100 iterations.

good tracking of the data by the model is observed. The outstanding speed of convergence of the SVF method should be noted. In particular it has only taken about 200 samples for the SVF to converge whereas the II technique has taken about 500 times more samples to converge to the same values.

Experimental Simulation

In an alternative way of investigating the accuracy and efficiency of the proposed identification techniques, the prospective electrical circuit equivalent of the ER valve was constructed using known dummy capacitors and resistors according to Figure 1. The circuit was excited in a similar fashion to the ER valve. Voltage and current time series data were captured and after non-dimensionalisation the identification procedures were applied. The data in these trials are closer to the ER valve data because all of the instrumentation, data capture, etc. (hence noise and inaccuracies) are similar in both cases.

Table 2 shows the results for a typical case when R = 54 kΩ and C = 19.6 nF. It is evident that as the excitation frequency is increased the accuracy in the determination of the resistance coefficient by both methods decreases slightly. As shown by the mathematical simulation the II approach requires around 100 iterations before convergence of the parameters, while the SVF only requires one iteration (i.e., 1000 sample points). A point which will be discussed further in the next section is that, at very low excitation frequency the II does not produce a sensible value for the capacitance even after 100 iterations. In contrast the SVF identifies the capacitance reasonably accurately. Both methods yield accurate values for the resistance parameter.

IDENTIFICATION OF EXPERIMENTAL DATA

Test Facilities and Procedure

A brief description of the experimental apparatus and procedure is given here. However, further details can be found in reference (Peel et al., 1988).

Table 2. Comparison between SVF and II using experimental simulation. True values: C = 19.6 nF, R = 54 kΩ.

Excitation Frequency	SVF Method		1st Iteration		II Method			
					50th Iteration		100th Iteration	
	C	R	C	R	C	R	C	R
2	25	52.8	520.8	47.6	253.9	49.9	200.1	50.5
60	19.5	50.6	29.4	51.1	21.5	51.5	20.9	51.6
200	19.3	44.6	20.1	50.2	18.9	47.8	18.8	47.7

The valve test rig comprised a simple hydraulic circuit powered by a speed controlled gear pump. The pump drew fluid from a temperature controlled reservoir via a combined magnetic/mechanical filter and delivered it to the experimental valve. A positive displacement flowmeter and a heat exchanger formed the return to the reservoir. The valve itself was a double annular flow channel type, with a mean pitch circle diameter of 115 mm. It was 100 mm in length and a gap of 0.5 mm was maintained between the plates.

Tests were performed on a ER fluid which was a mixture of dielectric base oil with some 30% by volume of lithium polymethacrylate solids. The porous solid contained approximately 10% water. The bias voltage across the valve plates was supplied by a high tension unit and the sinusoidal perturbation (±50 volts amplitude), generated by a signal generator, was then superimposed on the bias voltage. Further particulars of the excitation instrumentation, current and voltage monitor circuits are given elsewhere (see Peel et al., 1988). Because of the nature of the ER fluid, i.e., being similar to a semi-conductor (hence rapid variation of its conductance with temperature change) and various other reasons the tests were limited to a temperature of 30°C ± 0.1°C.

Simultaneous recordings of four channels, i.e. voltage, current, valve pressure difference and pump speed (flow rate) were made possible by using a transient data capture system. The recorder had a 10-bit resolution and a sampling rate of up to 200 kHz was possible on all four channels. Analogue signals were passed through conditioning equipment before being recorded by the capture system. In particular, a low pass/brick wall filter was always set at half the sampling rate on each channel so as to eliminate false reading via anti-aliasing; the filter cut off frequency was variable from 0.1 Hz to 100 kHz. After the capturing process, all the data was transferred to an IBM main frame (3083) where the identification procedure was then carried out.

Results and Discussion

Although the tests were performed at various operating conditions, i.e., flow rates, bias voltage etc., two typical cases only are dealt with in detail in this paper. It is believed that these and other data presented in the simulation section will suffice the present aim in scrutinizing the methods of analysis of the data from the ER controller system. Further-

more, they establish the basis for specifying the type of data needed from a new shear mode apparatus (see Bullough et al., 1991) and more importantly enhance the confidence in the results to be attained from those data.

In a previous publication (Firoozian et al., 1989) it was asserted that there was little variation in the values of R and C as the flow rate was altered. Hence in the present paper only one flow rate (i.e., 15 ℓ/min) is considered. However, the effect of two values of bias voltage (i.e., 600 V and 1200 V) are studied in detail.

Figures 4 and 5 present the results of the identification procedures. As explained earlier, Equations (3) and (4) were employed to obtain values of R and C from the identified parameters a and b. Figure 4 demonstrates the variation of resistance with the excitation frequency, whereas Figures 5(a) and 5(b) depict the capacitance change as a function of frequency.

It is very clear from Figure 4 that the two identification techniques yield very similar resistance results, and in fact at high bias voltage away from very low excitation frequencies the two sets of results are almost identical. At lower bias voltage the trend is very similar, although there is a slight difference in the absolute values of the resistance. The mismatch at low frequencies will be discussed later.

In Figures 5(a) and 5(b) the capacitance results from the two identification methods are compared with each other. Again, as for the resistance values, the capacitance results are very much in agreement at high and moderate frequency values. Although they significantly differ at low frequencies, the trend is still the same for both identification techniques.

For accurate and complete identification of the parameters involved in any system, the input and output signals must be sufficiently rich. That is, they must contain adequate information about the system. In the present electrical modelling of the ER fluid there are two unknown parameters, i.e., the resistance and capacitance. Hence, at least two independent sources of experimental information should be available to determine these unknown parameters. Fortunately that is the case; the amplitude ratio of the voltage to current and the phase shift between these two signals provide the necessary information. However, if these information are insignificant (i.e., very small phase shift) and/or corrupted with large measurement noise (i.e., high noise to signal ratio) the accuracy in identifying the unknown parameters diminishes very significantly.

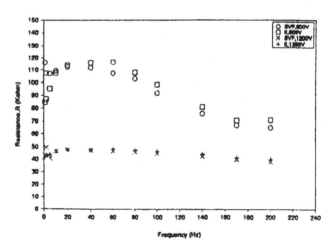

Figure 4. *Estimated resistance from biased sine wave excitation voltage, flow rate = 15 l/min. ○○○ SVF, bias voltage = 600 V; □□□ II, bias voltage = 600 V; × × × SVF, bias voltage = 1200 V; + + + II, bias voltage = 1200 V.*

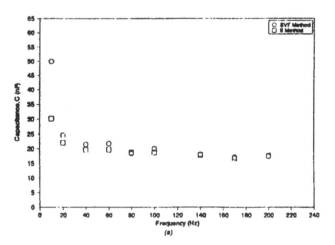

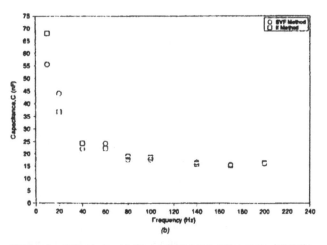

Figure 5. *Estimated capacitance from biased sine wave excitation voltage, flow rate = 15 l/min: (a) bias voltage = 600 V, (b) bias voltage = 1200 V, ○○○ SVF method results, □□□ II method results.*

The modelling of the ER fluid as is carried out in this paper and in (Firoozian et al., 1989; Peel et al., 1988) is a new idea, thus little information is available. Therefore, one must not overlook the possibility of choosing a wrong model structure. However, assuming the selected electrical RC network model is correct, due to the reasons explained above, identified results should be treated with some reservation at low frequencies.

One of the reasons for lack of confidence in the RC model is that the fitting of the estimated data to the experimental ones at low frequencies was not very good or at least as good as the results at high frequencies. This was despite the fact that on some occasions the parameters a and b seemed to converge to steady values. Therefore, one may argue that there might be other element(s) in the model which have not been included. These elements seem to have insignificant effect at high frequencies, whereas, their effect became more apparent as the excitation frequency is reduced. Another reason which supports the notion of missing elements in the model structure is that, at frequencies below 5 Hz a phase lag (instead of phase lead) is observed (see Figure 2). Originally it was deemed to be due to instrumentation equipment, however, careful examination showed that the phase lag was genuine and could only arise from the ER controller system.

One possible reason could be the existence of an inductive element in the system. A recent publication (Firoozian et al., 1990) has argued against such likelihood. However, in that paper only one particular arrangement of the RLC circuit has been studied and thus there remains several other ways in which these elements could be organised. Therefore, further careful studies are essential in order to increase the confidence in both the model and the estimated parameter values.

Since the phase angle between the voltage and current signals become very small at low frequencies and excessive measurement noise may corrupt the data substantially (see Figure 2), this could be regarded as another reason for the inconsistencies observed at low frequencies. In an attempt to artificially increase the phase shift, capacitors were connected in parallel with the ER fluid. Analyses of the data subsequently captured gave rise to the following observations: (a) The resistance values were unchanged. (b) The fitting of experimental data by the model improved. (c) The estimated ER fluid capacitance dropped substantially compared to the values determined in the absence of the additional capacitors.

CONCLUSIONS

Two methods of determining the resistance and capacitance values in a RC model of an ER fluid have been tried. Both simulated and experimental data have been used to compare the identification techniques. Although, the two parametric identification approaches were accurate, the computational efficiency of the SVF method was very much higher than that of the II technique.

Both methods produced similar results at high and moderate frequencies (i.e., between 40 Hz and 200 Hz), but wide discrepancies were observed at low frequencies.

NOTATION

a	Non-dimensional voltage coefficient ($= 1/RC\omega_e$)
b	Non-dimensional current coefficient ($= I_m/C\omega_e V_m$)
C	Equivalent electrical capacitance of the fluid
c	Known constant (filter coefficient)
$f(\ )$	A function of
I	Normalised current
i	Current supplied to the valve
J	Least squares cost function
N	Number of sample points
$\underline{P}$	Error covariance matrix
$\underline{R}$	Equivalent electrical resistance of the fluid
$\underline{S}^{-1}$	Data covariance matrix
t	Time
V	Normalised voltage
v	Voltage supplied to the valve
$\underline{y}^T$	Estimation filter state vector
x_1, x_2, x_3	State variables of the model in the II method
$y(t)$	Measured signal plus noise
z	Auxiliary variables used in SVF method
θ	Identified parameter vector
ξ	Measurement noise
τ	Non-dimensional time
HQH	A weighting matrix
(T)	Indicate the transpose of a matrix
$(\cdot)$	Derivation with respect to time
$(\sim)$	Indicate a vector or matrix
$(\times)$	Indicate estimate

REFERENCES

Bullough, W. A., 1989. "Miscellaneous Electro-Rheological Phenomena", *Proc. 2nd Int. Conf. on Electrorheological Fluids, Raleigh, NC.* Lancaster, PA: Technomic Publishing Co., Inc., pp. 115–157.

Bullough, W. A. et al. 1991. "The Electro-Rheological Catch/Latch/Clutch", *Proc. I. Mech. E. Eurotech Direct. ICC Birmingham*.

Detchmendy, D. M. and R. Sridhar. 1966. "Sequential Estimation of States and Parameters in Noisy, Non-Linear, Dynamic Systems", *Trans. ASME, J. Basic Eng.*, 88:362–368.

Ellis, J., J. B. Roberts and A. Hosseini Sianaki. 1988. "A Comparison of Identification Methods for Estimating Squeeze-Film Damper Coefficients", *ASME Journal of Tribology*, 110(1):119–127.

Firoozian, R., W. A. Bullough and D. J. Peel. 1989. "Time Domain Modelling of the Response of an Electro-Rheological Fluid in the Flow Mode", *Vibration Analysis—Techniques and Application DE-Vol. 18-4*, ASME, pp. 45–50.

Firoozian, R., D. J. Peel and W. A. Bullough. 1990. "Magnetic Effects in an Electro-Rheological Controller", *Proc. I. Mech. E. Mechatronics Conf., Cambridge*, pp. 231–238.

Gawthrop, P. J. 1984a. "Parametric Identification of Transient Signals", *IMA Journal of Mathematical Control and Information*, 1:117–128.

Gawthrop, P. J. 1984b. "Parameter Estimation from Noncontiguous Data", *IEE Proceedings*, Vol. 131, Part D, No. 6, pp. 261–266.

Gawthrop, P. J., A. Kountzeris and J. B. Roberts. 1988. "Parametric Identification of Non-Linear Ship Roll Motion from Forced Roll Data", *Journal of Ship Research, SNAME*, 32(2).

Peel, D. J., W. A. Bullough and R. Firoozian. 1988. "The Derivation of the Electrical Characteristics of an Electro-Rheological Fluid from Biased Sine Wave Tests", *Proc. 8th Int. Fluid Power Symp., NEC Birmingham*, pp. 527–545.

Peel, D. J. and W. A. Bullough. 1989. "Miscellaneous Electro-Rheological Phenomena", *Proc. 2nd Int. Conf. on Electroheological Fluids, Raleigh, NC*. Lancaster, PA: Technomic Publishing Co., Inc.

Roberts, J. B., J. Ellis and A. Hosseini Sianaki. 1990. "The Determination of Squeeze Film Dynamic Coefficients from Transient Two-Dimensional Experimental Data", *ASME Journal of Tribology*, pp. 288–298.

Young, P. C. 1984. *Recursive Estimation and Time Series Analysis*. Springer-Verlag.

Some Aspects of Electrorheology

E. V. KOROBKO

Luikov Heat and Mass Transfer Institute
Byelorussian Academy of Sciences 220728
Minsk

ABSTRACT: The present work is an attempt to give an insightful review of the available data on the electrorheological effect (ERE) and its nature under different strain conditions.

It covers the known publications dealing with ERE starting with the pioneering works of W. Winslow to the present-day manuscripts. Particular attention is paid to the research studies of Soviet scientists over the past 20 years. The main approaches to the consideration of an ERE mechanism are discussed, and a model is proposed to describe the behavior of high-concentration suspensions under small strains. The resulting measurements of the main rheological characteristics (ERE viscosity, plasticity, elasticity) are reported. Contributing external factors (i.e., temperature, magnitude, and type of shear forces, an electric intensity, as well as a suspension composition) on an ERE magnitude are elucidated. Also, consideration is given to the role of a time factor and the kinetics of structurization in the case of the electrorheological effect.

The results of experimental studies on the mechanical behavior of poor-conducting suspensions in the presence of electric fields have found embodiment in the original designs of different devices and arrangements for hydroautomatics, i.e., controllable valves, pumps, brake pushers, etc.

In addition, a series of arrangements is proposed to fasten samples, the surfaces of which are subjected to such mechanical treatment as precise sharpening, grinding, cutting, and other finishing procedures. The results are presented for the application of ERE technology in the gripping devices of robots and damping units.

INTRODUCTION

THE electrorheological effect was first discovered by an American engineer, W. Winslow, in 1942 (Winslow, 1949) and has served as a basis for developing a new trend in forming the junction between some branches of knowledge: physical-chemical mechanics, rheology, electrochemistry and electrophysics, and colloid chemistry. Not until the 1960s in the USA were specific features and a scope of application of silica-based electrorheological composites in dielectric dispersed media defined. However, the instability of their properties has not allowed a wide application of the main advantages of the electrorheological effect in industry, namely, direct influence on working compositions without intermediate transformations, reversibility of properties, high-response, and energy-saving as compared to known devices. Further research studies have been conducted in the USA, Great Britain, Japan and West Germany mainly on a contract basis with private and governmental organizations.

Since 1966 in the USSR purposeful and systematic research studies of electrorheological fluids (ERF) based on dispersed natural and polymer fillers have been carried out in the rheo-physics laboratory at the Heat and Mass Transfer Institute of the Byelorussian Academy of Sciences (Minsk, USSR). A dispersion medium is to be any low-viscosity hydrocarbon fluid with a high electrical resistivity (10^{16} to 10^{20} Ohm·m, $\epsilon = 2$ to 5). A dispersed phase must possess high adsorptivity, have a highly hydroxylated developed surface (200 to 400 m²/g is desirable). It includes silica (diatomite, aerosyl), aluminosilicates (clays, talc, mica), oxides of metals (titanium, aluminium, zinc), and starch. Used particles are sized from 1 to 10 μm. The third substance, an activator, was applied to the surface of a solid-state phase. Possible good activators are: first, organic compounds capable of forming hydrogen bonds (water, ethyl, butyl, propyl alcohols, ethylene glucol, diethylene glucol, etc.) and, second, amines (aliphatic, aromatic, primary, secondary). Their amount, as a rule, is experimentally chosen with respect to a maximum change of the effective viscosity in the presence of an electric field. Surfactants are used as the fourth component of an ERF. They allow for greater particle volume fractions to be utilized in the high-concentration ERF case. For lower particle volume fraction suspensions, surfactants increase sedimentation stability. Most often, slightly-soluble non-ionic surfactants are used, e.g., fat acids and their ethers with alcohols (glycerine, monooleate, sorbite oleates and stearates, etc.). The use of surfactants amounts to 1–3 molecules per 2 μm² area of particles.

All ERFs known at present have some disadvantages, with a main drawback being a low electrorheological sensitivity characterized by a coefficient, $K = \Delta\tau/\Delta E$. Many compositions change their characteristics with time, exhibit a residual effect, and may be used only in definite temperature and humidity ranges (Figure 1*), which is not applicable for some technologies.

The known compositions of ERFs are improved by changing the ratio of the known components; for instance, the

---

*The dependence la has been obtained by I. Bukovich (Minsk, USSR).

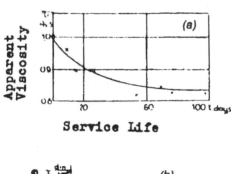

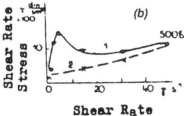

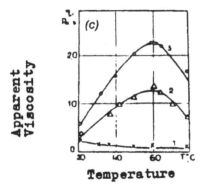

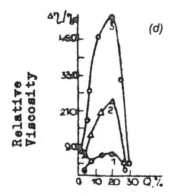

Figure 1. ERF typical properties: (a) ERF-DO apparent viscosity η (diatomite in a transformer oil, water being an activator) vs. time of use after its preparation. Rotatory viscosimetry data; (b) Pseudo-hysteresis: ERF shear stress vs. increasing (1) and decreasing (2) shear rates C = 40%, E = 0.66 kV/mm; (c) ERF apparent viscosity η vs. temperature. C = 1%, $\dot{\gamma}$ = 12 s⁻¹, E = 0(1), 1.5 (2), 2.5 (3) kV/mm; (d) Relative viscosity $\Delta\eta/\eta$ of 10% suspension of carboxy-methylcellulose in a transformer oil vs. moisture content at different electric intensities: E = 0.8(1), 1.6(2), 2.6(3) kV/mm.

amount of oleic acid in silica-base ERFs can be decreased (Gorodkin, Demidenko, and Novikova, 1980), or any component can be substituted for another belonging to the same class of materials as in the use of aviation oil instead of transformer oil as a dispersion medium (Shulman, Gorodkin, and Kuz'min, 1972). Attempts have been undertaken to improve fixing characteristics of an ERF composition, e.g., by their preliminary treatment at low temperatures (Shulman, Gorodkin, and Ragotner, 1988). Additional advantages without deterioration of the main characteristic of ERFs have been obtained by Kovganich, Fomenko, and Deinega (1979); and Blokh, Gelikman, and Gorodkin (1980). Thus, with the aim of expanding a working temperature range in the case of using a mineral powder in oil, Kovganovich, Fomenko, and Deinega (1979) have employed ferrites. In order to decrease thixotropic ERF properties, using a composite based on two solid fillers has been proposed, namely, bentonite clays and fine-dispersed silica (aerosyl).

All physical properties of electrorheological suspensions are stipulated by a structural factor. High-speed filming, direct visualization of the behavior of particles in a condenser gap, as well as spectro-interferometry have revealed that when an electric field is applied, particles of a solid phase uniformly distributed up to this moment in a fluid and randomly moving immediately cease their motion (Shulman and Matsepuro, 1974; Rubanov, Korobko, and Kitzan, 1991) and begin to oscillate from one electrode to another colloid and form associates, clusters and bridges, while in high-concentration ERFs they form a structural skeleton. Depending on conditions, the response time is from 10^{-4} to 10^2 s.

Some work has been done using a polymer solid phase in electrorheological suspensions (Shulman, Deinega, and Gorodkin, 1972). Thus, as a polymer filler, use has been made of phenol formaldehyde polymers (Stangroom, 1982), polymetacrylic and polyacrylic acids and their esters (Stangroom, 1980), polymer organic semiconductors (Block and Kelly, 1986), as well as cellulose and its derivatives (Uejima, 1972; Kordonsky, Korobko, and Lazareva, 1990; Sugimoto, 1985). In some cases polymers have contained acidic or other functional groups introduced by chemical modification or by adding low-molecular acids. It has been shown that one way to increase electrorheological sensitivity is by using polymer fillers not only in the H⁺-form but also in the ionic form of such salts as Li^+, K^+, Na^+, Cu^+, Al^{3+}, Cr^{3+}, Mg^{2+} (Stangroom, 1980).

THE ERE MAIN CONCEPTS

It should be noted that structural changes in the presence of an electric field (Figure 2) accompanied by the formation of individual aggregates are typical of other dispersions based on non-polar fluids and are well known (Vorobieva, 1967; Deinega and Vinogradov, 1984). However, unlike ERFs, they form structures with separate particles bound

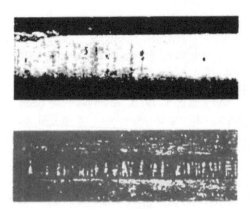

Figure 2. Photos of the ERF inner structure in a flat condenser gap at E = 0(1), 0.7(2), kV/mm, C = 2%.

weakly with each other and easily disintegrated even at a small load. Failure of an ERF structure is associated with the considerable consumption of a mechanical energy. Failure in real time depends on medium composition, electric force magnitude, and several external factors, such as temperature, pressure, and humidity.

The structurization process of ERF in the presence of an electric field is due to the polarization of components, conductance and coagulation (Usiyarov, Lavrova, and Efremov, 1966). As is known, the polarization of a dispersion medium is very weak and contributes little to the dielectric losses. The dielectric characteristics of a solid phase are important (Figure 3) though they do not always play the main role; experiments with ceramics such as BaTiO$_2$ (Seignette's salt having a dielectric constant of about 10^3) have demonstrated a weak electrorheological activity of ERF on their base (Kim, 1975). Estimates obtained by Shulman, Kulichikhin, and Dreval (1990) demonstrate that a level of the interaction energy of particles in structural units of ERF in the presence

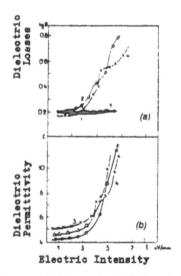

Figure 3. Influence of 50 Hz a.c. electric field on dielectric loss tangent (a) and dielectric constant (b) of air-dried cellulose (paper) samples (1) and carboxymethylcellulose with the different amount of COOH Groups ψ = 0.8(2), 1.4(3), 2.6(4)%.

of an electric field corresponds to the energy of intermolecular interaction. Obviously, the main type of ERF polarization is the interfacial polarization at the particle-medium interface attributable, as may be judged from large values of a tangent of dielectric loss to the displacement of free charges (Shulman, Deinega, and Gorodkin, 1972).

Different hypotheses exist concerning the contribution of the above factors to a structurization process. These hypotheses are partially supported by the experiments concerned with a study of the influence of an amount and type of activator, temperature and shear rate on dielectric characteristics (ε, tan δ, conductance) and an ERE magnitude.

Analyzing ERE in high-viscosity dielectric fluids, in which the displacement of a fine-dispersed solid phase is hindered in the presence of an electric field, D. Klass and T. Martinek have suggested that a change in the mechanical properties of suspensions is related to the deformation of double ionic layers (Klass and Martinek, 1967). This point of view is shared by Uejima adopting the Schwarz model (Schwarz, 1962) to describe polarization of spherical colloid particles with a counter ion medium (Uejima, 1972). In practice, in the majority of cases the sizes of particles in a dispersed phase are larger than those of the colloidal particles, i.e., an influence of the double ionic layers on their behaviour cannot be the determining factor. A structure of the double layer at the complex interphase boundary of a four-component ERF is not clear as well. G. Petrzhik et al. consider these surfactants to be responsible for ERE (Petrzhik, Chertkova, and Traperznikov, 1980). It is common knowledge that many surfactants are capable of transforming into a liquid-crystalline state characterized by an increase in viscosity, a structure strength, and an increase of a dielectric constant of these substances.

It is supposed that such a liquid-crystalline state (mesophase) is formed by an electric field in a surfactant layer surrounding particles. However, this hypothesis has not been confirmed by the authors and only later Ermolenko, Shulman, and Lazareva, (1987) have supported experimentally this hypothesis for solid polymer-based ERFs. This mechanism, however, may be considered as the accompanying but not the main one, since many ERFs not containing surfactants exhibit a good electrorheological effect.

Based on the experimental data, according to which a maximum ERE has been observed in systems containing water (dried particles of a solid phase are not structurized), a hypothesis suggesting the existence of water bridges between particles has also been proposed (Stangroom, 1983; Luikov, Shulman, and Gorodkin, 1972). Later J. Stangroom described it as the "osmotic" hypothesis or "the liquid adhesive" theory. He has suggested (Stangroom, 1983) that inside the pores of the particles in a solid phase there are ions with marked mobility. These ions group water molecules around themselves ("free" water chemically unbound with a particle surface unlike silica-bound hydroxyl groups). When an electric field is applied, the ions start to move and transport water towards one end of particles joined by their opposite ends with the aid of a liquid interlayer which, in principle, is similar to dipole formation. Thus, for instance, water added to flour forms a tough dough, and the work of mechanical forces is aimed at overcoming the surface forces in a suspension. A great deal of data is available to support this theory including data obtained by researchers at the Luikov Heat and Mass Transfer Institute of the Byelorussian Academy of Sciences (Minsk, USSR). First, it has been shown that ERFs with a too large amount of water or any other "activator" are transformed into a thin paste and respond weakly to an applied electric field (Luikov, Shulman, and Gorodkin, 1972). Second, the ionized solid particles have produced more active ERFs as compared to those containing no ions—alginic acid-starch). Of importance in this case is the charge density of mobile ions. For instance, calcium or magnesium ions with a high charge density are strongly bound and cannot move even in a high-potential field. Ions having a low charge density cannot retain water sufficiently and an ERF exhibits a weak effect. This hypothesis is supported by ERE experiments with the salts of different metals, among which lithium salts have the highest activity decreasing to cesium in a salt series which has the lowest charge density. Third, it has been observed that the chemical properties of a dispersion medium, on which the surface forces at the interphase boundary with an activated particle depend, exert influence on the magnitude of the electrorheological effect.

Unlike the deformation theory of solvate shells, this approach makes it possible to combine and quantitatively estimate the resistance forces occurring between particles which respond not only to shearing but to stretching (compressing) action as well. However, it does not explain at all any accompanying current increase in structure and a change in dielectric ERF characteristics in the presence of an electric field. Moreover, it is almost solely based on the statistical analysis.

An attempt to relate mechanical changes with specific features of a charge transfer for activated suspensions has been made by A. D. Matsepuro (Shulman, Matsepuro, and Khusid, 1974; Shulman, Matsepuro, and Khusid, 1977). Based on the results of experiments with large model silica gel spheres (Shulman, Matsepuro, and Khusid, 1974), the particles of an ERF solid phase have been presumed to be conducting with a certain density of surface charge. In an electric field they are polarized and oscillate in a gap, transferring a charge from one electrode to another. The force of ERF particle-particle interaction has been determined by measuring the potential of a conducting sphere placed at a large distance from the condenser plates (dipole approximation) (Shulman, Matsepuro, and Khusid, 1977). The influence of other charges has not been taken into consideration in the calculations and comparing experiments has produced a difference in the shear stress by an order of magnitude. Such an approach does not seem to be fully sound since many investigators have demonstrated that the conduction current measured in real ERFs is more typical of poor dielectrics or semiconductors than of suspensions conducting through structure.

H. Block (Block and Kelly, 1987), continuing Yu. Deinega and N. Kovganich's studies of a frequency dependence of dielectric ERF properties (Kovganich, 1977) has paid attention to the formation of relaxation peaks in the frequency range $10-10^5$ Hz. The mechanisms responsible for such a high level of low-frequency ERF polarization are the result of the migration of mobile charges but not of dipole orientation. It has been suggested that the relaxation time of polarization is a critical characteristic in the ER mechanism. It depends on the co-direction of an induced dipole moment on the particles swirled in shear flow and on an electric intensity vector. The mechanical work is concerned with the recovery of torque $[P \times E]$ and specifies to what extent a viscosity increases in the ERF flow with no marked structuring. Using these hypotheses, H. Block was the first to propose the employment of the unactivated particles with an electron-type conduction and a high electrorheological activity for ERFs (Block, Kelly, and Qin, 1989). However, this theory of the mechanism does not explain the resistance to the tensile-compressive force in ERFs and requires quantitative estimation and comparison with experiment.

Believing that polarization leads to Coulomb interactions, the majority of scientists try to describe them using different mathematical models with a larger or smaller degree of confidence (Deinega and Shilov, 1986; Adriani and Gast, 1988). In order to take into account the mutual effect of charges, nonsphericity of particles, a rate of polarization, as well as electrical phenomena on a double electrical layer surface, the dipole-dipole interaction scheme is used (Dukhin and Estrela-Liopis, 1979).

Thus Adriani and Gast (1988) suppose that suspended particles polarized in an external electric field interact as point dipoles induced by the external field. The force of their interaction is inversely proportional to the cube of the distance between particles. In this case, the motion of particles under the action of electric and hydrodynamic forces becomes correlated, creating high-frequency elasticity and viscosity of a suspension. A correct bridge structure in such a model does not develop and, as a rule, the system is correctly calculated for low-concentration ERFs. Extending the results formally to the case of the existence of the formed bridge structures without considering the repulsive forces between particles has not yielded real estimates and has not explained the saturation effect of electrorheological properties of suspensions. Some investigators (Dukhin and Estrela-Liopis, 1985; Shilov and Estrela-Liopis, 1979) have demonstrated the ERE magnitude by calculating only the multipole polarization interaction force of the central pair of particles in a complete bridge, which has also produced underestimated results. The best approximation is the calculations performed by Estrela-Liopis and Dudkin (1987) who have applied the multipole diagram-analytical method (the quadrupole approximation). Complicated calculations, including time scanning, have allowed the authors to determine the polarization interaction forces of particles. A comparison with the friction force parallel to and normal to an electric intensity vector, and an evaluation of critical parameters of bridge failure (an angle of its displacement and length) for starch suspended in vaseline (Deinega and Estrela-Liopis, 1990) has demonstrated satisfactory agreement with the experimental data at small and moderate electric intensities (up to 2.5 kV/mm). By increasing an electric intensity, a considerable deviation has been observed because of nonlinear, with respect to an electric field, polarization and electric stabilization of particles (Simonova and Shilov, 1978).

However, the electrorheological effect, as shown by many investigators, is a more complicated phenomenon associated not only with the polarization interaction of particles, but also with the absorption of charges at the dielectric-poor conductor interface at the expense of leakage current (the Jonsen-Rahbek effect) (Jonsen and Rahbek, 1923).

In this case the problem is reduced to a solution of the Poisson equation describing the occurence of charges on particles as a consequence of mutual induction. Considering the multipole potential factorization, the problem on determination of coefficients, even in the case of only two particles, results in an infinite set of differential equations and has no closed form solution.

In practice, other equivalent approaches to the solution of similar problems are adopted. Thus, in the theory of magnetic fluids (Rosensweig, 1989) the Brown theory on the equivalence of dipole and pole approaches is employed (Brown, 1951). The latter is based on the presentation of the interaction forces of two dielectrics in terms of the tension forces and pressure at their interface (Panovsky and Philips, 1918; Panovsky and Philips, 1963). The presentation of the interaction forces of polarized particles in terms of the tension force was first adopted in ERF considerations in Deinega and Shilov, 1986. However, calculations have led to a wrong conclusion, i.e., the independence of forces at a distance between particles, and this approach has not been extended. An error is associated with the one-sided consideration disregarding pressure forces in a dispersion medium layer between particles. In addition, no consideration has been given to spatial polarization of a surface of dielectric particles covered with an activator (water, as a rule) or changing their dielectric properties in a strong electric field.

Based on the "pole" approach as well as using the model similar to the known kinetic Frenkel liquid model (Frenkel, 1975), it is possible to build a model of a concentrated electrorheological fluid as a packet of solid dielectric plates with dielectric liquid interlayers cut along the planes normal to electrodes (Mokeev, Korobko, and Vedernikova, 1991). In such models the particles form a regular spatial network (skeleton) of the bridges of interacting particles distorted by hydrodynamic shear forces. In concentrated suspensions the distances between particles are considerably less than their sizes. In a first approximation, an electric field in an ERF

layer is uniform along the electrodes, while in a transverse direction (along bridges) it undergoes jumps at the boundaries between cubic particles and the thin interlayers of a dispersion medium between them. The cubic representation of particles, being of an irregular form in practice, is closer to reality because it takes into consideration the small thickness of an interlayer as compared to their size.

Prior to applying an electric field, an activator, depending on the amount, may form an envelope covering two or more particles. In the presence of an electric field, its surface (allowing for a small dielectric constant of a medium $\epsilon_m \leq \epsilon_p$) is affected by the Maxwellian tension force (Landau and Lifshits, 1982):

$$F = \oint [\vec{D}(\vec{E} \cdot \vec{n}) - \frac{1}{2}(\vec{D} \cdot \vec{E})\vec{n}]\, dS \qquad (1)$$

In the case of the electric field intensity parallel to the normal to a surface of particles, n, it is directed towards a phase with a smaller dielectric constant (oil). As a result, particles of a solid phase are attracted to each other with the force:

$$F = \frac{1}{2}\epsilon_o\epsilon_m \left(\frac{\epsilon_p}{\epsilon_m}\vec{E}_m\right)^2 S\vec{n} \qquad (2)$$

On a side surface of particles, where $\vec{E}$ is normal to $\vec{n}$, the force of the Maxwell tensions is directed towards large dielectric constants (compressive stress) (Landau and Lifshits, 1982; Tareev, 1982). In an inhomogeneous field there occurs a stress between the particles on the edge of a disperse interlayer which is the greater field gradient and the smaller interlayer thickness. Together with the forces occurring in a solvate envelope (a double electric layer), a force of this stress at some minimal interlayer thickness $\xi = d$ compensates the attractive force between particles. In the same manner, as is known, coalescence of water drops in an oil medium is prevented in the presence of an electric field (they only form chains from an electrode to an electrode but do not merge). With increasing ξ, the resultant force is that of attraction. It increases up to the value in Equation (2) and then, because of a decrease of the electric intensity gradient on an interlayer edge, enhancement of the edge effect, and weakening of a median field in a gap due to intensity redistribution between activator and oil layers in the suspension, the interaction force of particles begins to decrease and a bridge fails. Thus, the expression (2) gives the ultimate strength of a bridge.

With an account of adsorption processes, the total electric intensity E_m in an interlayer between particles is determined as

$$E_m = \left\{ \frac{\epsilon_p}{\epsilon_m}[1 - \exp(-t/T_i)] \right.$$
$$\left. + \frac{R}{\xi}[1 - \exp(-t/T_c)] \right\}E \qquad (3)$$

where

     T_i is the time of condenser charging
     T_c is the time of adsorption charge accumulation on a
        particle
     R is the mean particle radius
     ξ is the distance between particles
  ϵ_p, ϵ_m is the dielectric constant of particles and a medium,
     respectively

Since the time of variation of the strength limit of a structure is much larger than T_i, in calculations of this value per area of bridges contacting electrodes (taking into account the weight concentration of a solid phase) the following expression is used:

$$\tau = F/S$$

$$= 0.6 \cdot \frac{1}{2} \epsilon_o \epsilon_m \left\{ \frac{\epsilon_p}{\epsilon_m} + \frac{R}{d} [1 - \exp(-t/T_c)] \right\}^2 E^2$$

$$(4)$$

Calculations at $E = 2$ kV/mm give τ equal to 50 kPa which agrees satisfactorily with experiment (Gordon, Ragotner and Uutma, 1980).

THE SPECIFIC FEATURES OF ERF MECHANICAL BEHAVIOR

Soviet researchers (Shulman, Khusid and Korobko, 1987; Shulman, Yanovsky, and Korobko, 1989; Shulman and Korobko, 1975) and other scientists (Brooks, 1989; Conrad and Sprecher, 1987) have thoroughly studied the mechanism of ERF behavior under different deformation conditions for Poiseulle flow, Couette shear, and sinusoidal vibration impact. Consider a typical ERF, namely, the suspension of a finely ground diatomite in a transformer oil, with water being an activator and oleic acid serving as a stabilizer. In measurements, an electric intensity has been widely varied at room temperature in two traditional strain regimes, i.e., continuous shear strain for steady-state flows, and periodic sinusoidal-law shear strain (a rotational viscosimeter and a vibrorheometer, respectively). In the latter case, investigations have been carried out both in the linear strain region, i.e., at extremely small strain amplitudes not disturbing the internal structure of a system, and in the nonlinear one. The linear strain region has been controlled by the known methods. ERF rheograms are presented in Figure 4. The data has been recorded for fully developed shear flow conditions. As is easily seen, the higher the solid phase concentration, the greater the effective viscosity in the absence of the field (curves 1 through 3) and the anomaly of a viscoelastic behavior, when an electric field is applied (curves 4 through 8) [Figure 4(a)].

It may be noted that at $C < 5\%$ ($C = 5\%$ may be assumed as the threshold concentration leading to system

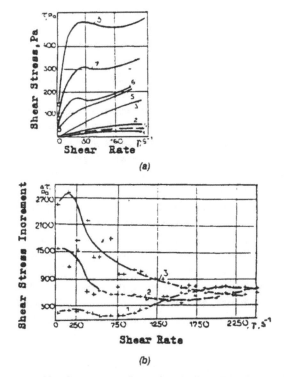

Figure 4. *ERF flow curves: (a) at $E = 0$, $C = 5(1)$, $20(2)$, $40(3)$; $E = 1.3$ kV/mm, $C = 5(4)$, $20(5)$; $E = 0.52(6)$, $0.67(7)$, $1.0(8)$ kV/mm, $C = 40\%$; (b) for ERF with 60% diatomite content in transformer oil at different electric intensities: $E = 0.5(1)$, $1.0(2)$, $1.5(3)$ kV/mm.*

structurization), a flow develops mainly in a liquid phase, adsorption layers formed on a surface of solid particles are not involved, i.e., the system behaves as a Newtonian liquid.

Addition of a finely dispersed silica filler to an ERF in the amount of more than 5% essentially changes the flow pattern of the medium, with an electric field being applied (curves 5–8). In these cases an ERF acquires, first of all a pseudoplasticity property described by the rheological equation of state $\tau^{1/n} \approx \tau_o^{1/n} + (\mu_p \gamma)^{1/m}$ where n, m, μ_p, τ_o are the functions of E. If in the absence of an electric field the dependence $\tau(\gamma)$ is linear and passes through the origin of the coordinates, then with the field being applied, there appears the yield limit τ_o, the values of which increase with an electric intensity, thus supporting the assumption about an increase of interparticle interaction forces and, therefore, strengthening of the system skeleton. The quantity τ_o is a function of concentration. Moreover, a change in electric intensity leads to the manifestation of thixotropic behavior. Irreversibility of the shear stress–shear rate curve results as the shear rate is varied in a cyclic fashion. Initial material structure is recovered when the ERF is exposed to a steady shear situation following cyclic shear rate variations. This recovery occurs more rapidly when low levels of electric field are applied. In the case of high-concentration ERFs at a transition from low to high shear rates, maximum appears on the $\tau(\gamma)$ curves. For such systems at high shear rates the $\tau(\gamma)$ curves qualitatively change their form; a decrease of τ values is observed with increasing γ, i.e., a decrease of an

ERF plastic viscosity. However, values of the apparent $\eta = \tau/\gamma$ [Figure 4(b)] exceed a zero-field viscosity even in the field of maximum shear rates $\gamma \approx 2500\ s^{-1}$. This value is indicative of the considerable contribution of electric field forces to deceleration of even separate ERF particles not connected through a skeleton.

The quasielastic-to-fluid state transition of an ERF depends on a load, a magnitude and action of an electric field (Figures 5 and 6). The dependencies presented show that the ERF transition from the state with an unfailed structure to shear flow, τ_o, is a variable parameter dependent on these factors and determined by the dynamic state of a system. Data on delay in yield development with a load applied (an induction period) have allowed evaluation of an activation energy from the equation of the kinetic theory of a strength of solids. Thus, for a 40% ERF in the presence of an electric field with the mean intensity $E = 0.75$ kV/mm, it amounts to 80 kJ/mol and approaches the energy of intermolecular bonds. Absolute values of the ultimate strength of an ERF structure which, unlike the yield limit, characterize discontinuity of a continuous phase, have been determined both at tangential displacement and their vertical separation from each other (Figure 7). Use has been made of solid or coplanar electrodes of a base. It is shown that the dependence of a shear resistance on E to its saturation is practically linear and depends on the type of base, its dimensions, roughness, thickness of protective coatings (Shulman, Gelikman, and Korobko, 1983; Shulman, 1975), and suspension properties. As a rule, a limiting force required for rupture of an ERF structure is an order of magnitude higher than forces necessary for its shear.

Evaluation of an ERF behavior at specific shears less than τ_o (a zone of low strains and quick recovery of a structure) has been performed by dynamometric methods at a sinusoidal load. It is seen (Figure 8) that in addition to viscosity variation there are recorded elasticity moduli G' as a function of electric intensity. The nonlinearity with increasing strain amplitude and the constancy of G' and loss modulus G'' in the frequency range 10^{-2}–10^2 Hz are observed. A comparison of the apparent viscosity values determined by the flow curves $\eta = \varrho/\gamma$, and dynamic viscosity $\eta^* = \left(\sqrt{G'^2 + G''^2}/i\omega\right)$ provided the equivalence of a shear rate and an oscillation frequency, shows that η^* values for ERF are higher than η in the entire test strain range (Figure 9). However, at small ω and γ these differences disappear. A different extent of failure of a structural skeleton is responsible for a discrepancy in τ and G'' values by two orders of magnitude. This descrepancy is attributed to specific features of a linear and nonlinear ERF behavior at different strain conditions.

A mechanism of an ERF behavior for the case of the Poiseuille flow (without a continuous structural skeleton) has been a matter of interest to many researchers (Brooks, 1989; Bullough and Stringer, 1973). Thus, it has been shown that if an electrorheological suspension is pumped through a channel-condenser, between the walls of which there is a

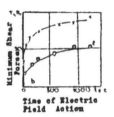

Figure 5. Minimum specific shearing force of ERF-DO in the presence of an electric field at $E = 1.5$ kV/mm (1) and after its switching off (2) at the moment of ERF shear vs. time of electric field action. $C = 20\%$.

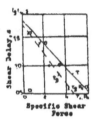

Figure 6. Surface life of ERF-DO ($C = 40\%$) preliminarily subjected to action of an electric field ($t_E = 60$ s) with $E = 0.75$ kV/mm vs. mechanical stress.

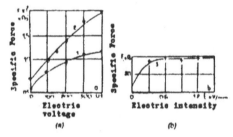

Figure 7. Specific separating (a) and shearing (b) forces of an aluminium sample vs. electric intensity for ERF-CO (clay in vaseline oil) and (1) $\in PC -D\ O \cdot C = 60\%$.

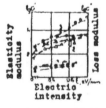

Figure 8. Log elasticity G' and losses G'' moduli vs. electric intensity at different diatomite concentrations: $C = 10(1)$, $20(2)$, $60(3)$ %; relative amplitude $\dot\gamma_o = 0.027$, $\omega = 0.628\ s^{-1}$.

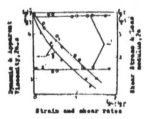

Figure 9. Viscoelastic characteristics obtained at periodic $[|\eta^*|$ (1), G'' (1')], and continuous $[\eta(2)$, $\tau(2)]$ deformation vs. frequency ω and shear rate $\dot\gamma \cdot C = 40\%$, $E = 0.65$ kV/mm.

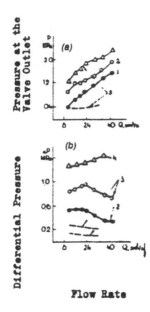

Figure 10. *Characterization of ERF flows via spiral and coaxial valves (a) flow rate vs. head; (b) differential pressure across a valve vs. flow rate E = 0(1), 2(2), 3(3), 4(4) kV/mm.*

sufficient potential difference, then a resistance to motion of this suspension may be regulated, depending on the electric intensity, until its apparent solidification, i.e., complete flow stop.

Results of experiments concerned with the measurement of flow rate characteristics for a flow channel in which a system of electrodes in the form of a coaxial-cylindrical or spiral value (Shulman and Korobko, 1975; Korobko, Mokeev, and Kolik, 1990) is embedded, are displayed in Figure 10. In the absence of an electric field, a pressure head linearly grows, i.e., the proportionality of the flow rate vs. pressure head curve is observed. With an electric field being applied, the flow rate vs. pressure head characteristic is nonlinear in the whole flow rate range (Figure 10). For instance, at $E = 2$ kV/mm it is nonlinear up to an inflection point to which a flow rate of 18 cm³/s corresponds; a section with a less intensive pressure rise follows ($Q = 18$ to 22 cm³/s). In the flow rate range 22–44 cm³/s the pressure increases more rapidly. At $E = 3$ kV/mm these regions are distinctly seen and displaced towards higher flow rates. The effective pressure drop, i.e., a portion of pressure head losses due to the electric effect on an ERF, has its maximum corresponding to a certain flow rate. As far as E increases, the maximum displaces towards higher Q values.

ER–DEVICES

The information obtained on ER-suspension properties has served as the basis for designing and bringing to a commercial level new, different, efficient, and inexpensive devices, apparatuses, and arrangements. Owing to the direct application of an electric field to a working medium (without intermediate transformations), mechanical, elec-

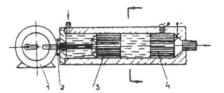

Figure 11. *A schematic diagram of a dielectric suspension pump (symbols are explained in the text).*

tromechanical and other analogous systems have been successfully replaced by simple electrorheological devices offering the convincing advantages of speed of response, service life, output characteristics (force, frequency), consumed energy, etc. The most known of such devices are hydraulic units as in hydroservo systems of machine tools (Shulman, Gordon, and Ragotner, 1979) automatic devices, as well as braking units, clutches (Shulman, Ragotner, and Korobko, 1979), or shock absorption systems adapting to external conditions. Consider the design and operation of some non-traditional ERF-based devices. Among these is a pump for transferring a dielectric suspension shown schematically in Figure 11.

A cylinder houses two valves 3,4. The suction valve 3 may perform reciprocating motion induced by drive 1, while the delivery valve 4 is rigidly fixed in the cylinder and maintains pressure in the pumping system.

When the suction valve moves towards the delivery valve, the necessary electric voltage 2 is applied to the former valve to provide a maximum increase in the apparent ERF viscosity up to locking. Serving as a piston, the valve expels the fluid, being in an intervalve space, via the delivery valve into the system.

When the suction valve achieves a dead point, electric voltage is applied to the delivery valve which prevents a pressure drop in the system, while the suction valve is deenergized. During a back stroke of the piston, the working fluid runs through the suction valve occupying an intervalve volume. Suspension pumping proceeds via repetition of the described cycles. Variation of the applied voltage allows continuous control of the flow rate and pressure in the system.

ER-valves are also used in electrohydraulic brake pushers (Figure 12) (Shulman, Gorodkin, and Blokh, 1978).

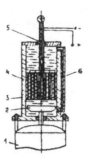

Figure 12. *A schematic diagram of an electrohydraulic brake pusher (for symbols see the text).*

Motor 1, rotating rotor wheel 2, induces a suspension flow to run through cylinder 3, electrorheological valve 4 and pumping-over channel 6 to the cylinder inlet beneath the rotor wheel. When an electric field is applied, the valve, operating as a piston, displaces together with rod 5 rigidly connected to it. The rod transfers the force to the brake linkage. The brake may be easily and quickly operated by de-energizing the valve or by energizing it, when it is necessary. Response to an electric signal does not exceed 0.05 sec.

The plot of ΔP on an electric intensity for a test system represents a family of almost straight lines with the center at the origin of coordinates. At high-intensity electric fields, an increase of solid phase percentage in a suspension exerts a more pronounced influence on the electrorheological effect and therefore, force characteristics of a device (Figure 13); for $E = 6.6$ kV/cm concentration variation of diatomite by 20%, e.g., from 30 to 50%, results in a 1.5 kg/cm² increment of ΔP, while at 20 kV/cm the increment is as large as three.

The effectiveness of ERF application has been confirmed by the model of a viscous-friction damper in a passive vibro-protection system. Figure 14 presents a comparison of performance characteristics under natural oscillation conditions. An effective decrease (5–6 fold) of a natural oscillation amplitude of an object with the weight $m = 200$ kg is seen when an electric field is applied. Resonance characteristics for different electric intensities (forced oscillations) are presented in Figure 15. A drastic increase of X is seen on the object until its coincidence with the inlet amplitude X_1, i.e., it is easy to attain the zero displacement of the object in the coordinate system of a base with the aid of external forces. Tests of the effectiveness of this method in the vibroprotection system of an electron microscope column ($m = 500$ kg) have produced positive results (Anaskin, Shulman, and Korobko, 1986).

Electrorheological technology applied to fixation of workpieces under mechanical treatment of their surfaces has also proven effective.

An arrangement for fixture of flat monolithic workpieces subjected to finish processing is shown in Figure 16 (Korobko, Ragotner, and Gorodkin, 1990). Anodized disk 2, ERF layer 3, workpieces 4, guard cover 5 ensuring radial support of the workpieces are arranged on horizontal chuck 1 of a lathe. An electric field from power source 6 is induced between the anodized disc (a base) and the workpiece. When being processed, a flat sample is affected by forces directed along it and perpendicular to its surface on the side of a tool and a suspension. Under the given conditions of precise turning on the rotating chuck-base, a shearing force on the side of a cutting tool is insignificant (about 20 g), but at the expense of high speeds of rotation an appreciable centripetal force acts onto the sample in the direction from the chuck centre. Electrodes via a system of current collectors have been connected to the power source with a control unit having an especially designed circuit diagram.

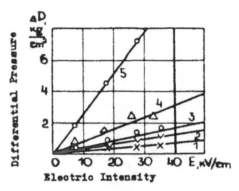

Figure 13. *Specific load on a brake pusher vs. electric intensity at different diatomite contents in ERF: C = 18(1), 25(2), 30(3), 35(4), 50(5) %.*

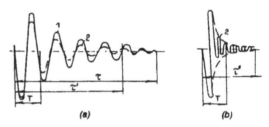

Figure 14. *Natural oscillation oscillograms of the sample with weight m = 200 kg on four supports with an ER-damper at E = 0(a) and 1.2(b) kV/mm. (1) fluid PMS-1000; (2) ERF-DO, C = 60%; r = 5 mm, K = 9.2 · 10⁴ N/m.*

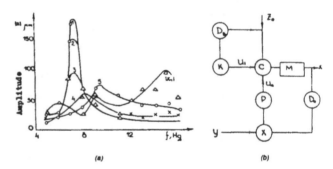

Figure 15. *The amplitude-frequency characteristics of the sample with m = 3.5 kg in the ER-damper system at different electric intensities. E = 0(1), 0.04(2); 0.08(3); 0.12(4); 0.3(5) kV/mm. K = 1.2 · 10⁴ N/m.*

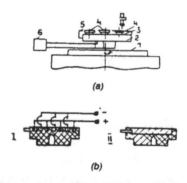

Figure 16. *Fixture scheme of flat workpieces with the aid of the ER-fixture arrangement at finish processing of their surfaces (a) and samples of discs-bases for fixing workpieces on a chuck (b).*

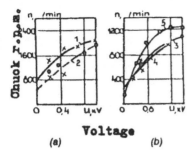

Figure 17. *The limiting number of chuck revolutions vs. electric field magnitude at separation (a) and shear (b) tests of workpieces with a weight of 108 g(a), 9.3g(b) with used solid (1) and planar (2–5) electrodes and antistatic varnish on the sample (3), the sample and electrodes (4). ERF(DO), C = 60%.*

The fixture arrangement has allowed determination both of horizontal and vertical displacement of the workpiece over the chuck surface. A speed of chuck rotation has been measured by a photoelectric sensor which has responded to a light beam passage through a hole in the disc mounted on the motor shaft.

Critical values of the rotation speed at which displacement of the workpieces takes place, i.e., the upper limit of the efficiency of the suggested method, have been determined for the worst case, namely, at the maximum distance from the chuck centre, i.e., at the maximum shearing force *F*. Flat circular workpieces (copper, aluminium) with the 8th-grade roughness of the base surface have been placed on a 50 μm thickness ERF layer and slightly pressed against the surface. The results of tests are presented in Figure 17. It has been established that the force of workpiece adherence to the rotating chuckbase is maximum when a solid electrode 11 is used (Figure 16). Planar electrodes have allowed critical values of a rotation speed to be obtained by 20 to 30% lower for the entire electric intensity range.

Mechanical treatment (precise sharpening) of workpieces fixed with the aid of an ER fixture arrangement has been performed under industrial conditions. A cutting depth has not exceeded 30 μm. At first, a workpiece face was processed by a steel cutting tool, while finish treatment was carried out by a diamond tool until the 14th-grade surface roughness was achieved.

Unlike the above conditions of processing the monolithic workpieces, the problem of processing the thin-wall, non-rigid constructions (envelopes, different reservoirs, panels, etc.) requires an account of the yielding of intermediate links (a lathe—an arrangement—a tool—a workpiece) and their sensitivity to elastic strains and vibrations.

The feasibility of increasing the precision of processing end face planes of thin-wall constructions has been experimentally studied on a setup shown in Figure 18(a) (Shulman, Korobko, and Gorodkin, 1991).

Construction 1 to be processed was a thin-wall shell of revolution made of an aluminium alloy and welded at butts. A height of the non-rigid construction was 1540 mm, a shell diameter was 2050 mm, a plate thickness amounted to 4

mm. Construction 1 was fastened on chuck 2 of a vertical lathe with the aid of strips 3. Electrorheological fixture units 4 coated with a 60% diatomite electrorheological suspension were brought into contact with the shell of construction 1 and fixed by clamps of racks 5 mounted on pedestals 6. The clamps are presented in Figure 18(b).

Each clamp consisted of aluminium-alloy electrodes with imbedded dielectric plates. Its bearing surface was processed following a surface contour of workpiece 1. The packet is fixed in cavity 6 of metallic base 5 with the aid of epoxy compound 3 serving simultaneously as insulation between the metallic base 2 and the electrodes 2. High voltage was applied to −ER-fixing units from power source 4 connected via a collector to the 220 V mains. When an electric field was applied to the fixture units, the effective hardness of an electrorheological fluid in a gap between the ER fixture device and the construction appreciably increased and workpiece 1 was fixed in the setup.

The absence of deformation of the thin-wall contruction during fixing and achieved uneveness of its face plane were controlled by an indicator set in a lathe slide.

In experimental works a 5 to 10-fold increase of precision (uneveness) of processing of end places is achieved as compared to the precision attained presently in industrial equipment in which its structural elements are used to fix a non-rigid construction.

The problem of optimal fixture of an object to be processed becomes more complicated if the object has a complex shape in addition to its non-rigid construction. For instance, when processing compressor blades (a modern motor with an axial compressor has up to 2000 blades), the requirements for precision and maintaining their shape are extremely high; only for a working section, i.e., a blade, the precision of manufacturing is 0.05 to 0.15 mm. Because expensive, high performance materials are used, conventional fixturing equipment considerably increases costs of production. For such workpieces the important requirement, in addition to those set for fixture means (low cost, universality, operativeness), is retention of rigidity of thin-wall shape manufactured articles.

The ER fixture constructions of a planar type considered above and not applicable in some technologies as, for instance, during processing of a rotatory blade on a grinding machine. A direct contact of a sublayer with its surface is re-

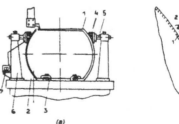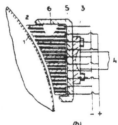

Figure 18. *Schematic of fixing a non-rigid shell-type construction on a chuck of a vertical lathe (a) and an electrorheological fixing unit for non-rigid cylindrical workpieces.*

quired and, in addition, a complex contour of a workpiece results in considerable costs for fabrication of special arrangements for each modification of the workpieces.

At the same time the possibility of reversing the fixing properties with the aid of an ERF in the presence of an electric field is rather valuable for the readjustment of equipment and for automatic systems of mechanical treatment on the whole. We have made an attempt to realize ERF advantages (high response, sufficient pressing force) in combination with the known principles of using the clamping devices having a point contact with the workpiece surface to be processed. There are mechanical arrangements in the form of stops (Tazemdinov, 1969) as well as hydromechanical constructions with clamping rods (Rogachev, 1977). In such arrangements rods displace at the expense of compression of a high-viscosity plunger, for instance, hydroplastic, by a plunger. A fixture force is determined as $F = PSN$ where P is the pressure, S is the rod area, N is the number of stops. The large number of used rods provides a uniform clamping force which approaches in value that of devices with a planar contact. The pressure in a working cylinder may be regulated only by direct application of a magnetic or an electric field to a ferromagnetic or electrorheological fluid without applying any additional mechanical force.

Based on the simple model of a hydrocylinder, a system to increase rigidity of blades has been designed which consists of a bracket with a fixed cowling inside of which ring 1 is mounted on movable supports.

The ring houses two sectors in inner cavities in which cylindrical bodies and movable rods 2 are rigidly fixed. The cavities are filled with electrorheological fluid 3. The voltage from a power source is supplied to the bodies and rods. The bracket is mounted on a guiding frame of a machine. A casing of the arrangement rotates on frames relative to the cowling (Figure 19).

Blade 4 to be processed is located between sectors of the arrangement body. Rods are pressed to its surface.

A specific feature of the designed arrangement is the feasibility of adapting the fixture elements (rods) to a blade tongue having spatial deflection.

When an electric field is applied to an ERF between a hydrocylinder and a spring-actuated rod, a cohesive force arises on their surfaces and there occurs blade reinforcement within the limits of the elastic strains of an ERF layer

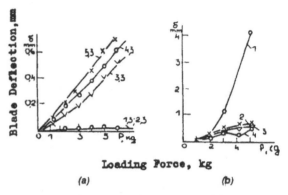

Loading Force, kg

(a)   (b)

Figure 20. *The influence of loading force (a) and fixture scheme (b) of a blade on its deflection at constant load. (1) fixture without tightening; (2) fixture with tightening; (3) with a support in an ER fixture arrangement without tightening; (4) fixture with a support in an ER fixture arrangement with tightening. 1.3 through 5.3, control points.*

or a blade itself. This is characterized by the ability of the whole construction to resist to external forces and to restore, when its action ceases, its initial parameters. As is known, on the processing of non-rigid workpieces a precision criterion is a magnitude of deviation of the object's shape caused by non-rigidity of the system "workpiece-arrangement" unlike the monolithic articles where the main criterion is a rough and defect-free surface of a workpiece.

With the aim of strain determination of a blade at different fixture schemes, experimental studies concerned with blade deflection determination at an applied calibrated force have been carried out. In the investigations the following fixture schemes for a blade root have been used: fixture without and with mechanical pressing; and with mechanical pressing and a support in an ER-arrangement in end and median sections.

The results of blade deflection measurements at control points depending on a fixture scheme and a load and position of a control point at an applied load from 1 to 6 kg are presented in Figure 20.

It has been established that an ER-arrangement, applied as a support, allows a decrease of blade deflections at the control points up to 30% which permits a considerable decrease of a dynamic component of a processing error.

The ER fixture technology may also be applied in robotics (Gorodkin, Gleb, and Korobko, 1984).

Transport of workpieces by automatic devices followed by their processing begins from the moment "to take", "to transport" to the moments "to put down" or "to install".

Effectiveness of displacement depends on a successful choice of a gripping device, this being one of the most important elements of a robot. The main requirements for gripping devices are as follows: reliability of gripping and holding a workpiece during displacement and at sudden stops; inadmissibility of damage of workpieces during gripping; easy interchangeability of gripping devices; minimal overall dimensions and weight.

Presently three types of gripping devices are used,

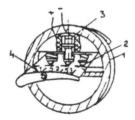

Figure 19. *Arrangement to increase rigidity of blades processed on a grinding machine: (1) supporting ring; (2) movable rods; (3) electrorheological fluid; (4) blade to be processed; (5.1–5.3) numbers of cross-sections.*

namely, mechanical, vacuum, and magnetic. Their preferable choice is specified by workpiece dimensions, material mass, and its continuity. Mechanical gripping devices are limited by a workpiece thickness. Vacuum gripping devices are restricted by continuity (no perforation) and a state of a material surface, while magnetic gripping devices are limited by material properties.

As experience has shown, a thin-sheet billet (workpiece) made of non-magnetic or dielectric material with through holes in the state of mounting or after preceding operations is not transportable by using the above gripping devices.

In small-scale and multiproduct-range production, a change of a gripping device requires much time, while manufacture of individual gripping devices for each type of workpiece increases the cycle of preparation procedures and requires considerable consumption of materials.

Depending on overall dimensions, design features, and mass of a workpiece to be displaced, complicated material-consuming gripping devices are often used. Their mass may be considerably higher than that of a workpiece to be displaced with restricts the load-carrying capacity of an industrial robot. Use of robots with an increased load-carrying capacity (and therefore more expensive) to compensate for the gripping mass lessens their technical-economical factors when being brought to a commercial level.

Therefore, the design of multipurpose gripping devices possessing minimal mass and providing the feasibility of gripping and transporting of flat and bulk workpieces is one of the urgent problems in the incorporation of robots into the many spheres of production and, particularly, sheet-stamping.

A schematic diagram of an ER gripping device is shown in Figure 21.

Electrodes 1–5 in the form of plates having a special coating are embedded flush with a base which is mounted to body 9. The base and the body are manufactured from a high-resistivity material. The body is connected with a holder which is fastened to an "arm" of a gripping device. A layer of an electrorheological suspension is applied to the base and the electrodes. When the electrorheological gripping device comes in contact with a workpiece, electrodes 1–5 are energized. At this moment the apparent viscosity of an electrorheological suspension between the working surfaces of a gripping device and a workpiece essentially increases and the workpiece "sticks" reliably to the gripping device. Depending on the workpiece material and mass, the

necessary composition of the suspension and magnitude of an applied potential are determined by which are sufficient to fix a workpiece in a gripping device, its withdrawal from a charging arrangement, transportation to a working zone of pressing or metal-cutting equipment, and its installation directly into a die or a chuck of a metal-cutting machine.

When a power source is switched off, the ER-suspension viscosity returns to its initial state, and a workpiece is separated from a gripping device which is backed away to repeat a cycle. Pivot gripping devices have been subjected to standard tests in a commercial-type robot system operating in a technological stamping process. The tests have involved workpiece withdrawal out of a charging arrangement, its delivery to a working zone of a press, its installation into a die, and its withdrawal after stamping and transferring to containers; 85 cycles have been conducted in an automatic regime. Workpieces made of magnetic and nonmagnetic materials have been taken from a charging arrangement in succession.

CONCLUSION

Unusual properties of electrorheological fluids allow for the development of non-traditional approaches to be applied to the problems of transfer and control of mechanical energy and the design of new devices. The mechanisms described above do not exhaust all the potentialities of ER-effect application. If the known restrictions are eliminated in the future, the ER-effect may be a promising tool in liquid electric generators, current transformers, electrokinetic balances, separation and purification apparatuses, in designing TV receivers with a plane television screen, acoustic devices (Shulman and Korobko, 1984), and different sensors and measuring systems on their basis (Korobko, 1991).

NOMENCLATURE

C	Concentration
d	Minimum interlayer thickness
$\vec{D}$	Electric displacement
$\vec{E}$	Electric field
$\vec{E}_m$	Electric field in carrier medium
F	Force
G'	Shear storage modulus
G''	Shear loss modulus
m	Mass
m, n	Constitutive law constants
N	Number of stops
$\vec{n}$	Unit normal vector
P	Pressure
P	Polarization
Q	Flow rate
r	Mean particle radius
S	Surface area

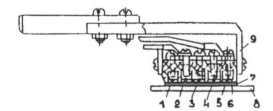

Figure 21. *A schematic diagram of an ER gripping device: 1 through 5, electrodes; 7, ERF; 8, workpiece; 9, body.*

T_i	Condenser charging time
T^e	Adsorption charge accumulation time
t	Time
$\dot{\gamma}$	Shear rate
δ	Dielectric characteristic angle
ϵ	Relative dielectric constant
ϵ_0	Universal dielectric constant
η	Apparent viscosity
η^*	Dynamic viscosity
μ_p	Plastic viscosity
ξ	Interlayer thickness
τ	Stress
τ_0	Yield stress
ω	Angular frequency

REFERENCES

Adriani, P. M. and A. P. Gast. 1988. "A Microscopic Model of Electrorheology", *Phys. Fluids*, 31(10):2757–2768.

Anaskin, I. F., B. K. Gleb and E. V. Korobko. 1984. "The External Electric Field Effect on Amplitude-Frequency Characteristics of an Electrorheological Damper", *Journal of Engineering Physics (Russian)*, 46(2):309–315.

Ballough, W. A. and J. D. Stringer. 1977. "Direct Flow Control by Electric Fields", *Hydraulic and Air Engineering*, Sept., pp. 11–14.

Blinkov, G. N., N. A. Fomin and E. V. Korobko. 1989. "Diagnostics of Field-Induced Structural Changes in Disperse Electrosensitive Systems", *Book of Abstracts of 5th Intern. Symp. on Flow Visualization*, Ang., Prague.

Bloch, G. M., B. F. Gelikman and R. G. Gorodkin. 1980. "A Composite Responding to an Electric Field", Author's Certificate (USSR), No. 1390235.

Bloch, G. M. and E. V. Korobko. 1978. "A Pump for Transferring Dielectric Suspensions", Author's Certificate (USSR), No. 606001.

Block, H., E. H. Gregson and A. Qin. 1983. "A Couette OLI with Fixed Startor Alignment for the Measurement of Flow Modified Permittivity and Electroviscosity", *Journal Phys. E. Sci. Instrument*, 16:896–902.

Block, H. and J. P. Kelly. 1986. "Electro-Rheological Fluids", UK Patent No. 2170510 A.

Block, H. and J. P. Kelly. 1987. "Electro-Rheology", *J. Phys. Desingh.*, 21(12):1661–1677.

Brooks, D. 1989. "Fluids Jet Tough", *Physics World*, Arg., p. 35.

Brown, W. F. 1951. "Electric and Magnetic Forces: A Direct Calculation", *Amer. J. Phys.*, 195:290–304.

Conrad, H. and A. F. Sprecher. 1987. "Characteristics of ER Fluids", *Proc. Adv. Materials. Conf. TMS*, p. 63.

Deinega, Yu. F. and V. R. Estrela-Liopis. 1990. "To the Theory of the Electrorheological Effect", *Abstracts of the XVth All-Union Symposium on Rheology, Odessa, USSR*, p. 64.

Deinega, Yu. F. and V. N. Shilov. 1986. "Structurization of Dispersion Systems in the Presence of Electric Fields", *Rheology of Polymer Systems and Rheophysics*, 2:56–64, Minsk, USSR: ITMO AH BSSR, Press.

Deinega, Yu. F. and G. V. Vinogradov. 1984. "Electric Fields in the Rheology of Disperse Systems", *Rheol. Acta.*, 23:636–651.

Dukhin, S. S. and V. R. Estrela-Liopis. 1985. "Electric Surface Phenomena and Electrofiltering", Kiev, USSR.

Ermolenko, I. N., Z. P. Shulman and I. G. Lazareva. 1987. "Electrorheological Effect of Titanium Dioxide Suspended in a Polymer Binder", *Dokl. Akad. Nauk BSSR.*, 31(10):906–909.

Estrela-Liopis, V. R. and V. V. Dudnik. 1987. "The Multipole Theory of Electric Conductivity of High-Concentration Suspensions of Charged Colloid Particles", *Colloid Journal (Russian)*, 49(1):110–115.

Ferry, J. D. 1961. "Viscoelastic Properties of Polymers", New York–London.

Frenkel, Ya. 1975. "Kinetic Theory of Fluids", Leningrad, Nauka.

Gordon, B. I., M. M. Ragotner and T. Kh. Uutma. 1980. "A Study of ERF Elastic Conditions under Static Loading Conditions", *Heat and Mass Transfer: Physical Fundamental Principles and Research Methods*, Minsk, USSR: ITMO AN BSSR Press, pp. 67–68.

Gorodkin, R. G., T. G. Demidenko and Z. A. Novikova. 1990. "Electrorheological Suspension", Authors Certificate (USSR), No. 1566712.

Gorodkin, R. G., V. K. Gleb and E. V. Korobko. 1984. "A Load-Gripping Device", *Abstracts of the Third All-Union Conference on Robot Systems, Voronezh, USSR*, pp. 51–52.

Jonsen, A. and H. Rahbek. 1923. *J. IEE.*, 61:713–714.

Kim, A. A. 1975. "Electrorheological Properties of Segnetosuspensions", *Rheology of Polymer and Dispersed Systems and Rheophysics*, Minsk, USSR: ITMO AN BSSR Press, pp. 88–96.

Klass, D. H. and T. W. Martinek. 1967. "Electroviscous Fluids, Rheological Properties", 38(1):75–79.

Kordonsky, V. I., E. V. Korobko and T. G. Lazareva. In press. "Electrorheological Polymer-Base Suspension", *Journal of Rheology*.

Korobko, E. V., A. A. Mokeev and V. L. Rolin. 1990. "Investigation of Performance Characteristics of Electrorheological Valves", *Proceedings of Int. Tech.-Transfer Congress "Actuator-90", Bremen*, pp. 191–195.

Korobko, E. V., M. M. Ragotner and R. G. Gorodkin. 1990b. "Electrorheological Workpiece Fixture Equipment", *Journal of Automation and Mechanization of Production*, 4:16–19.

Korobko, V. B. 1991. "Electrorheological Fluids as a Basis for Various Sensors of New Generation", *Proceedings 5th Intern. Congress for Sensor and Systems Technology, Nurenberg, May*.

Kovganich, N. Ya., Ya. F. Deinega and K. V. Popko. 1977. "An Influence of Electrical Frequency on Increasing a Viscosity of Dispersed Systems", *Ukrainian Chemical Journal (Russian)*, 43(12):1237–1240.

Kovganich, N. Ya., E. B. Fomenko and Yu. F. Deinega. 1979. "A Working Medium for an ER Fixture Arrangement", Author's Certificate (USSR), No. 688509.

Landau, L. D. and E. M. Lifshits. 1957. "Electrodynamics of Continua", Moscow: GITTL.

Luikov, A. V., Z. P. Shulman and R. G. Gorodkin. 1972. "An Influence of an Activator and a Nouisothermal Factor on ERF", *Proceedings of the IV All-Union Conference on Heat and Mass Transfer*, ITMO AN BSSR Press, 3:239–243.

Mokeev, A. A., E. V. Korobko and L. G. Vedernikova In press. "Structural Viscosity of Electrorheological Fluids", *Journal of Non-Newtonian Fluids*.

Panovsky, V. and H. Philips. 1963. "Classical Electrodynamics", Moscow: GIFML.

Petzhik, G. G., O. A. Hertkova and A. A. Trapeznikov. 1980. "Electrorheological Effects in Non-Aqueous Different-Composition Dispersion Media in Dependence on Electric Fields Parameters", *Dokl. Akad. Nauk., SSSR*, 1:173–177.

Rogachev, V. B. 1977. "A Device for Installing Complicated Geometry Workpieces", Author's Certificate (USSR), No. 553083.

Rosensweig, R. F. 1989. "Ferrohydrodynamics (Russian)", Moscow.

Rubanov, A. S., E. V. Korobko and A. I. Kitsan. 1991. "Evaluation of ERF Structurization Kinetics by the Special-Interferometry Technique", *Proceedings of Intern. School-Seminar Rheophysics and Thermophysics of Nonequilibrium Systems, May, Minsk*, pp. 77–82.

Schwarz, G. 1962. "A Theory of the Low-Frequency Dielectric Dispersion of Colloidal Particles in Electrolyte Solution", *Journal Phys. Chem.*, 66(12):2636–2642.

Shilov, V. N. and V. R. Estrela-Liopis. 1979. "Multiple Theory of Electrocoagulation of Lyophobic Sols", Moscow: AN SSSR Press, pp. 35–41.

Shulman, Z. P. 1975. "Electrorheological Effect and Its Applications", Minsk: ITMO AN BSSR Press, pp. 1–57.

Shulman, Z. P., Yu. F. Deinega and R. G. Gorodkin. 1972. "Electrorheological Effect", Minsk: Nauka i Technika.

Shulman, Z. P., B. Yu. Gelikman and E. V. Korobko. 1983. "The Method of Roughness Control of an Electro-Conducting Surface and a Device for Its Implementation", Author's Certificate (USSR), No. 1013748.

Shulman, Z. P., R. G. Gorodkin and G. M. Bloch. 1978. "An Electrorheological Brake Pusher", Author's Certificate (USSR), No. 625073.

Shulman, Z. P., R. G. Gorodkin and V. A. Kuz'Min. 1972. "A Working Substance for Dielectric Motors", Author's Certificate (USSR), No. 336755.

Shulman, Z. P., R. G. Gorodkin and M. M. Ragotner. 1979a. "A Servomechanism for Profiling Machines", Author's Certificate (USSR), No. 709329.

Shulman, Z. P., R. G. Gorodkin and M. M. Ragotner. 1988. "The Method to Obtain Diatomite Suspensions", Author's Certificate (USSR), No. 1404515.

Shulman, Z. P., B. M. Khusid and E. V. Korobko. 1987. "Damping Mechanical Systems Oscillations by a Non-Newtonian Fluid with Electric Field", *Journal of Non-Newton, Fluids*, 25:239–247.

Shulman, Z. P. and E. V. Korobko. 1978. "Convective Heat Transfer of Dielectric Suspensions in Coaxial Cylindrical Channels, Heat and Mass Transfer", No. 5, pp. 543–548.

Shulman, Z. P. and E. V. Korobko. 1984. "An Acoustic Lens", Author's Certificate (USSR), No. 1122374.

Shulman, Z. P., E. V. Korobko and R. G. Gorodkin. 1991. "Electrorheological Fixture Arrangements", *Preprint No. 8*, Minsk, USSR: ITMO AN BSSR Press.

Shulman, Z. P., V. G. Kulitchikhin and V. E. Dreval. 1990. "Thixotropic Properties of ERF under Continuous Straining", *Journal of Engineering Physics (Russian)*, 59(6):34–40.

Shulman, Z. P., A. D. Matsepuro and B. M. Khusid. 1974b. "Charge Transfer by Oscillating Particles in an ERF. Experimental Studies", *Izvest. AN BSSR, Ser. Fiz-Energ. Nauk*, 4:62–69.

Shulman, Z. P., A. D. Matsepuro and B. M. Khusid. 1977a. "ERF Structurization in the Presence of an Electric Field I. Qualitative Considerations", *Izvest. AN BSSR, Ser Fiz-Energ. Nauk*, 3:116–122.

Shulman, Z. P., A. D. Matsepuro and B. M. Khusid. 1977b. "ERF Structurization in the Presence of an Electric Field, II. Quantitative Considerations", *Izvest. AN BSSR, Ser. Fiz-Energ. Nauk*, 3:123–129.

Shulman, Z. P., A. D. Matsepuro and M. B. Smolsky. 1974a. "A Study of ERE Structural Characteristics in Non-aqueous Dispersed Systems", *Izvest. AN BSSR, Ser. Fiz-Energ. Nauk*, No. 1:60–66.

Shulman, Z. P., M. M. Ragotner and E. V. Korobko. 1979b. "A Clutch with Electric Control", Author's Certificate (USSR), No. 684211.

Shulman, Z. P., Yu. G. Yanovsky and E. V. Korobko. 1989. "The Mechanism of a Visco-Elastic Behavior of Electro-Rheological Suspensions", *Journal of Non-Newt. Fluids*, 33:181–196.

Simonova, T. S. and V. N. Shilov. 1978. "The Electrostabilization Effect of Dispersions in a Stationary Electric Field", *Colloid Journal*, 40(1):81–87.

Stangroom, G. 1982. "Imprelectrik Field Responsive Fluids", UK Patent No. 2100240.

Stangroom, G. 1980. "Improvements in or Relating to Electric Field Responsive Fluids", UK Patent No. 1570234.

Stangroom, G. 1983. "Electrorheological Fluids", *Journal Phys. Technol.*, 14:290–296.

Stevens, N. G., I. L. Sproston and R. Stanway. 1984. "Experimental Evaluation of a Simple Electroviscous Damper", *Journal of Electrostatic*, 15:275–283.

Sugimoto, N. 1985. "Application of Winsloss Effect to the New Actuators", *Journ. Japan Society of Lubrication Engineers*, 30(12):859–864.

Tareev, B. M. 1982. "Physics of Dielectric Materials", Moscow: Energoizdat.

Tazetdinov, M. M. 1969. "Electric Fixture Arrangements", *Machines and Tools (Russian)*, 3:33–34.

Uejima, H. 1972. "Dielectric Mechanism and Rheological Properties of Electro-Fluids", *Jap. Journal of Appl. Phys.*, 11(3):319–326.

Usiyarov, O. G., I. S. Lavrova and I. F. Efremov. 1966. "On Contribution of Polarization Interaction to Electrophoretic Deposition", *Journal of Colloid. Chem.*, 28(3):596–603.

Vorobieva, T. A. 1967. "On Manufacture of Fibrous Structures from Polymer Dispersions in the Presence of an Alternating Electric Field", *Problems of Physical-Chemical Mechanics of Fibrous and Porous Disperse Structures and Materials*, Riga: Zinatne Publishing House, pp. 23–25.

Material Aspects of Electrorheological Systems

KEITH D. WEISS* AND J. DAVID CARLSON
Thomas Lord Research Center
Lord Corporation
Cary, NC 27511

JOHN P. COULTER
Department of Mechanical Engineering and Mechanics
Lehigh University
Bethlehem, PA 18015

ABSTRACT: This paper provides a summary of the current state of electrorheological (ER) material research and development. In particular, a description of the electrorheological effect, a definition of observed behavior, a critique of the proposed mechanisms for the ER phenomenon, correlations among the properties exhibited by ER materials, an overview of the various test methodologies for characterizing ER materials, and a discussion of the requirements imposed by specific applications is presented. Inherent throughout this paper are references to the properties exhibited by currently available, state-of-the-art ER materials. In order to facilitate the design of devices and to evaluate the effectiveness of a particular ER material in a specific application, it is necessary that experimental data reported for each material be consistent in terminology and test methodology. A recommendation as to the minimum amount of mechanical/electrical property information needed to adequately evaluate an ER material is provided.

INTRODUCTION

THE ability to control the rheology of a material with an applied electric field is of interest to both industrial and academic communities. The development of theories and models to explain this electrorheological (ER) phenomenon has been fueled by the overall market potential for this technology. The multi-million dollar sales predicted for electrorheological materials have been overshadowed only by the potential market value for the devices utilizing these materials [1,2]. Although such published market studies are perhaps overly optimistic, the simplicity of engineering designs based on ER material technology has facilitated the development of specifications for a broad range of devices, such as dampers, clutches and adaptive structures. The primary barriers to establishing a commercial ER business have been the lack of satisfactory materials and an inadequate understanding of the ER phenomenon.

Three articles published prior to 1990 provide detailed descriptions of the ER phenomenon in terms of electrical/structural variables [3–5], the status of theoretical understanding [3–5] and material composition [4]. Overviews of possible applications also are included in two of these articles [4,5]. Although these articles provide an excellent guide to the field of electrorheology, a tremendous amount of research and development has been conducted in this area over the past two years. It is not our intention to duplicate this previously published work, but rather to use it as a basis on which to build. This survey consists of a pair of articles covering both ER materials and potential applications. This first paper includes a description of the electrorheological effect, a definition of observed behavior, and a critique of the proposed mechanisms for this phenomenon, as well as a correlation between the properties exhibited by electrorheological materials, the various test methodologies used for characterizing these materials, and the requirements imposed by specific applications. The utilization of the controllable mechanical behavior exhibited by ER materials in the design, construction, and modeling of various system components such as valves, clutches, mounts and brakes, as well as in intelligent structures, is presented in the second paper [6]. It should be noted that throughout these two papers, we refer to electrorheological fluids or suspensions as electrorheological materials. One should not confuse these electroactive materials with the structural materials used in the construction of devices.

THE ELECTRORHEOLOGICAL PHENOMENON

A small reversible change in the viscosity of glycerin, castor oil and heavy paraffin within an electric field applied transverse to fluid flow was first observed by A. W. Duff in 1896 [7]. Subsequent investigations of this phenomenon have shown that a visco-electric effect arises in conductive fluids that either contain ionic impurities [8,9] or permanent dipoles [10,11]. Several mechanisms for this phenomena, such as electrophoresis [11,12], molecular orientation [10,11], induced dipole moments [11] and space/charge effects [11,13,

---

*Author to whom correspondence should be addressed.

14] have been proposed. The weak nature of this visco-electric effect in isotropic liquids has limited their utilization in any practical applications. The visco-electric effect described above should not be confused with the electrorheological phenomenon observed for particle suspensions in a fluid.

The formation of pearl-chains of neutral particles suspended in a medium differing from the particles in dielectric character was first observed by Priestley [15] and Winckler [16] in the 18th century. The occurrence of this phenomenon has since been observed in a variety of biologically related systems [17–20]. Further quantitative investigations of this pearl-chain formation under both DC and AC electric fields [21–23] have resulted in considerable theoretical speculation [24,25]. The formation of dipoles on non-spherical particles in an electric field will result in mutual dielectrophoresis and dipole-dipole attractions. The observed alignment of particles parallel with an applied electric field is believed to be related to the displacements and torques produced in the medium by the electric field and the translational motion and relocation of particles to positions having minimum potential energy [26].

The observation of a large induced electrorheological effect was first reported by W. Winslow in 1947 [27]. It was not until he published the results of his investigation on the Johnsen-Rahbeck Effect in 1949 that the field of electrorheology was born [28]. His initial disclosure described the electrically induced fibrillation of small spherical dielectric particles suspended in low viscosity oils [29]. Many of the fluid compositions and theoretical mechanisms originally described by Winslow are still relevant to ER materials under evaluation today. The demonstration that the rheology of a flowable viscous suspension could be controlled within a millisecond time frame through the application of an electric field inspired much thought concerning potential applications for this technology, such as brakes, dampers, clutches and hydraulic equipment.

Definition of Observed Behavior

The electrorheological effect was initially defined as the apparent change in viscosity observed in the materials developed by Winslow. Although from a macroscopic point of view, a change in apparent or effective viscosity does occur, the actual plastic viscosity (η) of the material defined as the change in stress per unit change in shear strain rate remains approximately constant as the applied electric field is varied. In this situation, the parameter that changes is the amount of shear stress needed to initiate flow. Many variables, including temperature, pressure, shear strain rate, and concentration, are important considerations in describing the flow behavior of ER materials. The classical Newtonian definition of apparent viscosity as the linear relationship between shear stress (τ) and shear strain rate ($\dot{\gamma}$) has been used by many authors to describe the ER effect at a specific electric field strength. However, ER materials exhibit approximate Newtonian type flow characteristics only in the absence of an electric field. An example of the typical shear stress versus shear strain rate behavior observed for an ER material in the presence of an electric field is shown in Figure 1. Currently, no recognized flow phenotype completely models or describes the observed behavior for all ER materials. However, a Bingham plastic model, as described by Equation (1), can often provide a sufficiently accurate description of the observed behavior to be used for the designing of ER material devices.

$$\tau_y = \tau(E) + \eta\dot{\gamma} \tag{1}$$

The Bingham plastic model recognizes that the property of an ER material generally observed to change with an increase in electric field is the yield stress defining the onset of flow. The electric field induced yield stress, $\tau(E)$, and viscosity, η, are the two most significant parameters used in designing electroactive devices where flow properties or post-yield properties are essential [30–33]. The dynamic yield stress ($\tau_{y,d}$) in a Bingham plastic modeled ER material can be defined as the zero-rate intercept of the linear regression curve-fit. The static yield stress ($\tau_{y,s}$) is defined as the stress necessary to initiate flow within the ER material regardless of whether or not a Bingham model accurately describes the material's behavior. Naturally, the plastic viscosity of the material in the post-yield regime is accurately reflected by the slope of the linear regression curve fit used in the analysis.

Many scientists [4] have reported observing a higher static yield stress than dynamic yield stress as shown by Curve A in Figure 1. The actual cause of this phenomenon, which is known as stiction, is not completely understood. It is suggested that this apparent static yield stress is a mechanical or structural anomaly of ER materials which is highly dependent upon particle size and particle shape, as well as the prior electric field and flow history of the material. In designing a device to utilize a particular ER material, it is necessary to consider the possible occurrence of this stiction. Upon returning from the flow regime to the

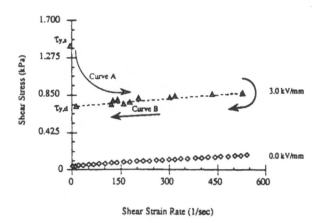

Figure 1. *Shear stress versus strain rate data obtained for Lord ERF/6533-30B.*

static situation, the rheology for an ER material is observed to follow the more typical behavior exemplified by Curve B in Figure 1. Several authors have attempted to correlate this occurrence with the time required to form favored fibrils or aggregates within the ER material [34]. However, it is suggested that the observation of this more typical behavior is a result of either the limitations encountered with the test methodology, electrophoresis, or electrode polarization.

Since ER materials consist of conductive particles surrounded by a dielectric medium, they essentially function as leaky capacitors. It is the leakage of charge between particles that results in the measurement of an electric current through the ER material. The charge flow is represented by an average current density, J, measured in units of current per unit electrode area. The current density associated with a particular ER material is useful for estimating the power consumption of devices using this material [30].

One property of ER materials that has not been extensively studied is the time required for the components of the material to respond to an electric field. In order for an ER material to be utilized in continuously variable, controllable devices, a response time on the order of a few milliseconds is required. In devices that are designed for on/off applications, it is essential to know the time required to reach the maximum control (shear stress) ratio or rheological equilibrium. One estimate of this response time can be obtained by measuring the rate at which the pressure increases in a constant flow rheometer when a sudden voltage is applied [35]. The single time constant response, T_s, associated with this flow measurement is taken to be the amount of time necessary for pressure to increase from a voltage-off baseline pressure to 63% of the final, voltage-on equilibrium pressure. The overall "turn-on" time for an ER material is generally taken to be 2–3 time constants. This assumes that the system behaves as a first order system and that the pressure rise when the fluid is subjected to a step input voltage can be written as shown in Equation (2), where p_{on} represents the system pressure at rheological equilibrium, t represents the amount of time that voltage has been applied, and T_s represents the time constant.

$$p(t) = p_{on}(1 - e^{-t/T_s}) \qquad (2)$$

In continuously variable devices, the establishment of a final rheological equilibrium is important only upon initial system start-up. It is more important in the design of these devices to know how fast the components of an ER material can respond to an incremental change in applied voltage. It is possible to obtain an estimate of this incremental response time through the measurement of the ER material's dielectric spectra [35–38]. The relaxation time, T_r, associated with the polarization decay of an ER material, is determined from the frequency at which either the dielectric loss or loss tangent data goes through a maximum, or at the inflection point observed in the relative permittivity data. Verification of this measured relaxation time is easily achievable through

the utilization of arc diagrams and a corresponding test of linearity [35,39,40]. This measured relaxation time corresponds to the lower limit in time necessary for an ER material to respond to the application of an electric field.

In addition to a post-yield regime, the shear stress/shear strain behavior observed for ER materials also contains a pre-yield regime [41–43] as shown in Figure 2. It is unfortunate that this regime has been the subject of only a limited number of investigations [41–52]. This pre-yield regime is defined by a yield strain, γ_y, and a static yield stress ($\tau_{y,s}$), as well as a complex shear modulus, G^*, that is dependent upon electric field strength. The complex shear modulus can be separated into its real, G', and imaginary G'', parts, called the storage modulus and loss modulus, respectively. The loss factor, $\tan \delta_G$, can be obtained through the relationship G''/G'. In the design of non-flowing devices, such as flexible or adaptive structures, control of the pre-yield complex shear modulus is essential.

The ability to design a practical device utilizing the electrorheological effect requires an accurate knowledge of the mechanical and electrical properties for the chosen ER material. In order to facilitate the design of devices and to evaluate the effectiveness of a given ER material in a specific application, it is necessary that the experimental data reported for each composition be consistent in terminology and test methodology. The following list describes the minimum amount of information that is needed by an engineer to adequately evaluate a particular ER material within a specific device design:

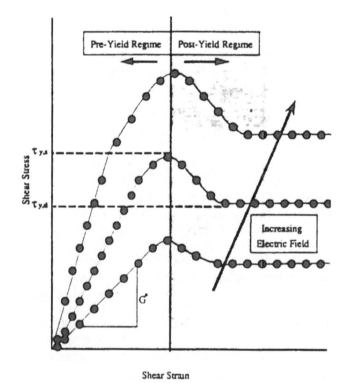

Figure 2. Figurative description of pre-yield and post-yield regimes observed in the shear stress/shear strain behavior of an ER material.

- shear stress versus shear strain data over the appropriate range of applied electric fields
- dynamic and static yield stress values (if Bingham plastic behavior is observed) as a function of applied electric field)
- zero-field viscosity
- response time estimate
- current density as a function of shear rate and electric field
- pre-yield complex shear modulus as a function of both applied electric field and frequency

For most applications, the measurement of these material properties over a broad temperature range is essential. The absence of this information in the published literature drastically hinders both the initial device design and its subsequent optimization.

Proposed Mechanisms for the ER Effect

It is generally accepted that the electrorheological phenomenon originates from particle polarization induced by an electric field. Although many models [5,46,53–70] for the interaction of an electric field with particles suspended in a fluid under shear have been proposed, no general consensus regarding the mechanism for the observed ER phenomenon has been reached. This is unfortunate since a thorough understanding of the mechanism is necessary to adequately optimize fluid compositions for specific applications. Attempts to validate these models through correlation with experimental data have only recently been initiated [57,66,70]. The proposed theoretical interpretations for the interparticle interactions that occur in ER materials, which will be discussed in the following subsections, include a distorted double layer, water "glue", and particle polarization.

Prior to discussing possible mechanisms for the ER phenomenon, it is important to comprehend the physical restructuring of the ER material components that occurs when the material is exposed to an electric field. Winslow's early work demonstrated the formation of a fibrous mass when particles suspended in low viscosity oils were exposed to an electric field [29]. This induced particle fibrillation is similar to the particle orientation observed to occur for charged colloidal particles [71–74], for silicon particles in a water medium [75], and in several biological systems [76]. Winslow suggested that the mutual attraction of spherical particles in regions of high electric field intensity leads to the formation of particle chains between electrodes. Winslow observed that the induced fibrillation in ER materials was similar to the magnetically induced fibrillation of iron suspensions [77]. Many other research scientists subsequently have attributed the observed behavior of various particle suspensions within a continuous DC, pulsatory DC, or an AC electric field to this fibrillation [53,54,78–93]. Microscopy studies [29,79,80,94] have provided evidence for the existence of these fibrillated structures as shown in Figure 3 for Lord ERF/115.

Electric Field →

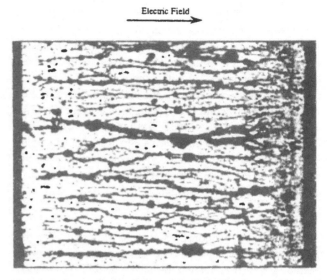

Figure 3. *Photomicrograph of particle fibrillation observed for Lord ERF/115.*

When dielectric particles having a permittivity different than that of the fluid in which they are suspended are subjected to a uniform electric field, the particles become polarized and the local field is distorted. The migration of these polarized particles to areas with greatest field intensity, a phenomenon called dielectrophoresis by Pohl [95], causes the interelectrode fibrillated structures to form [55]. These particle chains are held together by interparticle forces and can have sufficient strength to inhibit or even prevent the fluid from flowing. In the presence of a shear stress, the equilibrium that is established between the formation and breaking of the interelectrode chains [54,78,82] corresponds to the yield stress defined in the Bingham plastic model. When the electric field is removed, the particles return to a random distribution, allowing fluid flow to resume.

The primary criticism of particle fibrillation has been that the amount of time required for mass migration does not correlate with the millisecond response time observed for the electrorheological effect. It is this criticism that caused Klass and Martinek [36] to propose the distortion of a double-layer as an explanation for the ER phenomenon. Recently it has been reported that in very dilute suspensions subjected to small electric fields, particle motion can be observed for as long as 20 seconds [46]. This observation should not be confused with the particle fibrillation that occurs in typical ER materials that contain a high particle concentration. In these materials, the particles need move only a distance equivalent to a fraction of a particle's diameter, e.g., microns, in order to form chains. The question that needs to be answered is not so much whether particle fibrillation occurs, but rather, what happens to these particle chains under shear? Do short chains persist? How fast do broken chains reform? The rearrangement of the fibrillated mass at low shear rates through the breaking and reforming of fibers created by the interaction of particles has been observed in a recent investigation utilizing a flow visualization technique [54]. At a high level of strain, the formation of both a fluid and solid region occurred, with the fiber structure remaining intact within the solid region. Most of the concern over the formation of a single fibril structure or the lack of this structure in an ER material is academic from an application point of view. Utilization of ER materials in any practical device requires such a high loading of particles that individual chains are not discernable. In this case, the formation of thick columns or three-dimensional particle structuring is evident [66,67,96].

DOUBLE-LAYER THEORY

As previously mentioned, Klass and Martinek were the first to propose the induced polarization of the double layer surrounding each individual particle in the fluid as a plausible mechanism for the electrorheological phenomenon [36, 37]. This double layer can be defined as the asymmetric distribution of charges caused by the influence of an external potential. This theory is schematically depicted in Figure 4. In general, the application of an electric field causes movement of the double layer surrounding each particle in the direction of the electrode opposite in charge to the ions within the double layer. Since a double-layer representation of interfacial interactions, such as the potential difference between surfaces [97], permittivity at low frequencies [98–100] and non-ohmic conductance [101,102] is a well-known concept, many authors have proposed this mechanism as being the primary polarization method observed in ER materials [38,46,78,89,92,98,103–105]. This double-layer mechanism is different than a typical coulombic interaction between dipoles. In this instance, the distortion of the double layer surrounding the particles in the presence of an electric field leads to a non-equilibrium charge distribution within the double layer, resulting in charge repulsion and/or

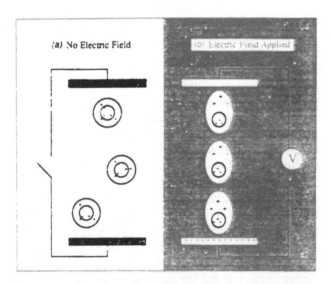

Figure 4. Schematic representation of double-layer theory: (a) no electric field; (b) electric field applied.

attraction. The main criticism of this mechanism has been that the Debye length of the double layers surrounding the particles are too large. In fact, it can be argued that the extent of an individual double layer is often greater than the distance actually separating the electrodes. This inconsistency has led to the emergence of several additional mechanistic interpretations.

WATER "GLUE"

Stangroom has proposed that water acts as a bridge between particles through an electro-osmosis process [79,106, 107]. He has suggested that the minimum requirements needed by a material in order to exhibit an electrorheological response include the presence of a hydrophobic liquid, a hydrophilic and porous particle, and an appreciable amount of water adsorbed on the particle. This mechanism assumes that ions trapped within the pores of the particle increase their mobility by dissolving in the water. In the presence of an electric field, these mobile ions shift the water layer surrounding the particle toward the oppositely charged electrode. Thus, one side of the particle becomes overly saturated with water. Overlap of the water between the particles is considered to be the bridge or glue that causes the ER effect. Upon removal of the electric field, the water and ions return to the pores within the particle, resulting in the dissipation of the induced dipole. Although it is possible that this mechanism is active in systems inherently containing water, the development of ER materials containing semiconducting particles [108], lithium hydrazinium sulfate [109], and substantially "anhydrous" alumino-silicates [110], has eliminated the possibility of the presence of water being a necessary prerequisite for the observation of the ER effect.

PARTICLE POLARIZATION

Several authors have proposed a third mechanism, particle polarization, to explain the same aspects of the ER phenomenon as previously described in both the double-layer theory and the water bridge concept [4,78,82,92,108,111]. There are five recognized modes through which polarization of the particles in an ER material can occur [112]. The most common of these modes, called electronic polarization, arises from the small distortions that occur within the positive and negative charge distribution of atoms. Although the contribution of electronic polarization to the overall polarization of the particle is small, it can directly influence the particle's dielectric constant. Another contributing factor to the dielectric constant of the particle is the movement within the solid lattice of charged atoms when they are exposed to an electric field. This type of polarization, called atomic polarization, contributes to the dielectric constant of inorganic particles to a larger extent than for organic particles. The occurrence of a form of rotational polarization, called dipolar polarization, is possible if the particle contains atoms that have a permanent dipole. This form of particle orientation may play an important part in the polarization of macromolecules. A fourth type of polarization mechanism, called nomadic polarization, occurs through the movement of thermally generated charges in a particle over several lattice interstices. This type of molecular polarization can involve either the movement of electrons ("hyperelectronic") or protons ("hyperprotonic"). It should be noted that the polarization effect induced by this nomadic mode can be quite large compared to both electronic or atomic polarization modes. The final mode of polarization arises from the piling up or accumulation of charge at the interfaces between particles, and is termed interfacial or Maxwell-Wagner polarization. In this mode, charges, either electronic or ionic, are free to migrate from one side of a particle to the other. Such a bulk movement of charge leads to the creation of enormous dipole moments. Charge migration may be through the actual bulk of the particle or along exterior or interior surfaces. For a highly porous particle having a large interior surface area, the distinction between volumetric and surface conduction is not particularly crucial. Interfacial polarization generally occurs in nonhomogeneous systems where the various materials have different conductivities and the charges are free to move. From a macroscopic perspective, interfacial polarization is difficult to distinguish from intrinsic electronic and atomic polarization.

It has been suggested that the origin of polarization [4], such as charge migration through the bulk, on the surface or in the double layer, does not matter. Instead, it is the rate and magnitude of this polarization that is important in the ER effect. In an ER system, it is the interparticle interactions that lead to the formation of particle clusters or aggregates. If a particle's rotation is too fast, i.e., field direction changes too rapidly, the effective polarization of the particle is small and the ER effect is diminished. Thus, an optimum rate of polarization will maximize the observed ER phenomenon. Once polarization has been established, the interaction of interparticle coulombic forces leads to the formation of a fibrillated network. It is assumed in this polarization theory that the fibrillated structures degrade in the presence of a flow regime. However, it is possible that aggregation or particle clustering can still occur under a flow regime within a limited lifetime. The occurrence of interparticle aggregation in ER fluid formulations containing high particle loadings would lead to the rotation of particle clusters instead of individual particles. It should be noted that this theory was initially presented by Mason to describe the effect of an electric field on very dilute, i.e., packing fraction < 0.02, suspensions [113]. The extrapolation of this theory to ER materials has been based upon flow-modified permittivity (FMP) studies involving polymer solutions containing various macromolecules [108,114].

Optimization of the ER Effect

The effects of various particle/fluid/additive components and process parameters on the macroscopic properties exhibited by ER materials, such as yield stress, current density

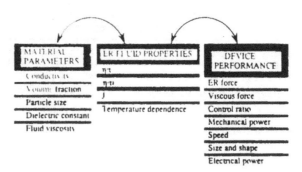

Figure 5. *Graphical representation of the possible interactions between material parameters, ER fluid properties and device performance.*

and stability, must be understood in order to develop materials suitable for practical devices [115–119]. The ability to control these variables to obtain the desired macroscopic ER material properties determines whether or not the desired device performance can be achieved. Calculations regarding the minimum levels for macroscopic properties needed by ER materials to perform within various device constraints have been previously discussed in the literature [30]. The major obstacle to commercialization of ER material technology has not been the availability of device designs, but rather, the practicality of these designs based upon the properties exhibited by available ER materials. A contributing problem is the lack of reproducible and comparable material test methodology.

An enormous number of process variables in different material systems have been shown to have varying effects upon the behavior and responses observed for the ER phenomenon (Figure 5). It is generally agreed that the current status of knowledge concerning these basic relationships includes an increase in the ER effect with higher particle volume fractions and higher field strengths, as well as a decrease in the proportionate effect with higher shear rates. Several authors have previously summarized the important requirements [4,33] and general specifications [33,34,120] for ER material properties. Although it is likely that the properties of a particular material will need to be tailored for a specific application, it is possible to identify the minimum property values that will allow an ER material to function in several broad, potential markets. These minimum ER material properties, which include yield stress, current density, zero-field viscosity, temperature dependence, and stability, are summarized in Table 1. In addition, material cost will also be an essential factor in the design of many devices.

EXISTING MATERIAL COMPOSITIONS AND PROPERTIES

Post-Yield Properties

In general, an ER material consists of a particle constitu-

Table 1. *Minimum properties needed by an ER material to be utilized in a variety of applications.*

Property Description	Suggested Value
Dynamic yield stress at 4.0 kV/mm	>3.0 kPa
Current density at 4.0 kV/mm (DC)	<10 μA/cm^2
Zero-field viscosity	0.1–0.3 Pa·s
Operating temperature range	−40 to +200°C
Dielectric breakdown strength	>5.0 kV/mm
Particle size	~10 μm
Stability	Low sedimentation No dynamic separation No electrophoresis No chemical changes Low volatility
Miscellaneous properties	Non-abrasive Non-toxic Non-corrosive Non-flammable

ent, a dielectric fluid component, a surfactant, and in some cases, an activator phase. The best method of evaluating the potential of utilizing various ER materials in a device design is to survey the patent literature for available material compositions with regard to their respective properties. A variety of material compositions, which have been patented since the initial discovery of the ER phenomenon in the late 1940s, are shown in Table 2. Unfortunately, the properties reported in the patent literature are vague, obtained by different test methodologies, inconsistent in definition, and varying in units. For instance, the strength of the electroviscous effect for material compositions has been defined in the patent literature as one or more of the following: (a) ratios of viscosity at a given electric field strength to the viscosity at zero field [129,132,143,159]; (b) ratios of static yield stress to a particular voltage [136,138,139,162]; (c) measured values of pressure [124–127,144]; (d) energy [145]; or (e) yield stresses derived from standard stress versus strain modeling [109,133,140,141,148]. Even in the various patents that define the strength of the electrorheological effect in a similar manner, the values usually are not directly comparable because the data was taken at either different electric field strengths, shear strain rates, or temperatures. In only one patent [148] are the actual shear stress versus shear strain rate curves at various electric field strengths provided for different material compositions. This lack of uniformity in reporting material properties has hindered the development of feasible device designs and the commercialization of ER materials.

YIELD STRESS MEASUREMENTS

Although there are inconsistencies in the methodology used to report material properties, it is possible to make some general correlations among the published data. It is recommended that any comparison of yield stress values reported for different material compositions be done with cau-

tion since the actual definition of yield stress used by various authors may not be consistent. If the yield stress values reported for these systems correspond to the dynamic yield stress as defined by the Bingham plastic model, a graphical comparison as shown in Figure 6 can be useful. Material formulations that exhibit dynamic yield stress values ranging from 100 Pa to greater than 3 kPa are currently available. It should be recognized that these dynamic yield stress values may not reflect the data that would be obtained after optimization of the reported formulations. In addition, this graphical comparison does not include material formulations that have been either studied at very low volume particle fractions, i.e., polyaniline suspensions in silicone oil [167], or reported with electrorheological activity defined in other terms, i.e., polyelectrolytes in paraffin oil [168].

Several authors prefer to describe the electrorheological effect in terms of the static yield stress [136,138,139,162]. In general, the static yield stress values reported in the literature are greater than the dynamic yield stress values described in Figure 6. For instance, an ER material consisting of 40% poly(naphthalene quinone) in a polychlorinated hydrocarbon has been reported to have a static yield stress at 2.5 kV/mm of approximately 7 kPa [139]. It is difficult with the available data to make any comparisons between reported static yield stress values and the dynamic properties of the various ER materials. As previously described in Figure 1, the occurrence of striction related behavior could cause the static yield stress to be much greater than the dynamic yield stress for a particular ER material. Furthermore, the dependence of striction on the prior electric field and flow history of the ER material establishes a level of uncertainty in the reproducibility of static measurements. In order to increase the controllability of a device based on ER material's post-yield behavior, it would be advantageous if the material were to exhibit similar static and dynamic properties.

ZERO-FIELD PROPERTIES

A comparison of yield stress values as a function of voltage alone does not provide sufficient information about the zero-field properties of the ER material. Knowledge of the zero-field viscosity and apparent yield stress associated with a particular ER material is essential because they are used in basic design equations [30], as well as in the definition of the device control ratio. Although Newtonian carrier oils are used in the preparation of ER materials, most ER formulations will exhibit a small zero-field yield stress. The occurrence of this apparent yield stress is highly dependent upon the loading, shape and degree of dispersion for the particles in the formulated ER material. The off-state properties of only a few of the ER materials described in Figure 6 are adequately reported in the literature. The zero-field properties of these ER materials are typically less than 200 Pa for the apparent yield stress and 0.35 Pa·s for the plastic viscosity. The most useful method of reporting the

Table 2. *Summary of material compositions reported in patent literature.*

Particle	Fluids	Surfactants	Additives	Ref. #
Starch, limestone, gypsum flour, gelatin, carbon	Transformer oil, olive oil, mineral oil	Not mentioned	Not mentioned	[27]
Silica gel	White oils, transformer oil, dibutyl sebacate, di-2-ethyl-hexyl adipate	Sorbitol sesquioleate, ferrous oleate, lead naphthenate, sodium oleate, sodium naph-thenate, polyoxyalkylene de-rivatized sorbitol oleate, borax, metallic hydrates	Water, glycerine, diethylene glycol	[121]
Silica gel	Mineral oil, kerosene	Aluminum distearate, alumi-num tristearate, lithium stearate, lithium ricinoleate, phenyl alpha naphthylamine, sorbitol sesquioleate, lauryl pyridinium chloride, borax, tin oxide	Water, ethylene glycol, mono-ethyl ether	[122]
Silica xerogel, barium tita-nate, magnesium silicate, charcoal, aluminum octoate, α-silica, aluminum oleate, colloidal silica, aluminum stearate, polystyrene carbox-ylic acid polymer, calcium stearate, colloidal kaolin clay, crystalline D-sorbitol, dimethyl hydontoin resin, flint, quartz, lauryl pyridinium chloride, lead oxide, lithium stearate, mannitol, micro-cel-C, micronized mica, mo-lecular sieves, nylon powder, onyx quartz, rottenstone, white bentonite, zinc stearate	Liquid oleaginous vehicles (dielectric constant = 2–5.5 and visc. < lubricating oil), silicone oils, petroleum based hydrocarbons, mineral oil, polyalkylene glycols, ali-phatic esters, fluorinated hy-drocarbons	Nonionic/anionic/cationic sur-factants, 4,4-bishydroxy-methyl-2-hepta-decenyl-2-oxazaline, aminosilanes, es-ters of polyhydric alcohols, fatty acids, naphthenic acids, resinic acids and salts, pri-mary amines, cresols	Water, alcohols, amines, hydroxy and polyhydroxy ali-phatic compounds, NaOH, KOH, calcium hydroxide	[123]
Silica (water saturated)	Nonpolar oleaginous vehicle, mineral oil	Glycerol monooleate, sorbi-tan sesquioleate, sodium di-alkylsulfosuccinate, hexyl ether alcohol, butyl cello-solve, octyl alcohol, dodecyl alcohol	Butylamine, hexylamine, ethanolamine, 2-amino-ethyl-amine, diethylamine, diiso-propylamine, water, dibutyl-amine, morpholine, diethanolamine, triethyl-amine, triethanolamine	[124]
Alumina, alumina/silica mix-tures	Nonpolar oleaginous vehicle with dielectric constant < 10, mineral oil	Sulfonates, sulfontated alco-hols, fatty acids, glycerol monooleate, sorbitol oleates, stearates, laurates, fatty al-cohols, anionic/cationic/non-ionic surfactants	Amines, acetic acid, water, 1-hydroxyethyl-2-heptadecyl imidazoline	[125]
All prior art, silica	All prior art, white oil, sili-cone oil, oleaginous hydro-carbons	Fatty acids, fatty amines, gly-cerol, glycol esters, glycerol monooleate	Water, alcohols, ethylene glycol, conductive metals (i.e., copper), N-aminoethyl-ethanolamine, 1-hydroxy-ethyl-2-heptadecyl imidazo-line	[126]
Silica with surface modified by esterification by glycerol monoesters of fatty acids	All prior art, nonpolar oleagi-nous vehicles with dielectric constant < 10, mineral oil	Anionic/cationic/nonionic sur-factants, glycerol monooleate	Water, amines, 1-hydroxy-ethyl-2-heptadecyl imidazo-line	[127]

(continued)

37

Table 2. (continued).

Particle	Fluids	Surfactants	Additives	Ref. #
Silicone ionomer reaction product of an amine functional diorganopolysiloxane and an acid (i.e., phosphoric, hydrochloric, nitric or sulfuric acid)	Electrically nonconducting liquid	Not mentioned	Not mentioned	[52]
All prior art, boron, silica aerogel	All prior art, mineral oil	All prior art, glycerol mono-oleate	Not mentioned	[128]
Pyrogenic silica	Carbon tetrachloride, xylene, silicone oil, fluorocarbon, mineral oil, olive oil, castor oil, cottonseed oil, linseed oil, kerosene	Not mentioned	Formic acid, lactic acid, malic acid, octanoic acid, pyruvic acid, aniline, ethylenediamine, phenylcyclohexylamine, triethanolamine, pyridine, ammonium hydroxide, acetamide, formamide, resorcinol, diethylene glycol, acetic acid, malonic acid, oxalic acid, methanol, trichloroacetic acid, ethylene glycol	[129]
Carbon black, all others	All fluids, metal soaps, nonpolar hydrocarbons, paraffin sulfates	Not mentioned	Soluble dye	[130]
Toners, dyes, phthalocyanine, copper phthalocyanine, flavanthrone, substituted quinacridone, diphenyl thiazole-anthraquinone, methoxyphenyl-imidoperylene, cadmium sulfide/zinc sulfide phosphor, zinc oxide, diethyl carbocyanine iodide, phosphototungstomolybdic acid	All prior art	Not mentioned	Actinic radiation	[131]
Electrolyte modified starch	All prior art, dielectric oil	Sorbitan monooleate	Water, ammonium ion, metal chlorides, sulfates, acetates, or fluorides	[132]
Microcrystalline cellulose	Hydraulic oil, liquid paraffin, silicone oil, dibutyl sebacate, oleic acid, orthochlorotoluene	Not mentioned	Water, formamide, alcohol, NaCl, LiCl, magnesium chloride, aluminum sulfate, KCl, ammonium chloride, sodium acetate	[133]
Alginic acid, Sephadex ion exchange resin, polyvinyl alcohols, sorbitol, mannitol, monosaccharides, polyhydric alcohols	Trifluorovinyl chloride, polychlorinated biphenyl, xylene, P-chlorotoluene, fluorolube, Aroclor, toluene, carbon tetrachloride, chloroform, nitrobenzene, chlorobenzene	Sorbitan monooleate, sorbitan sesquioleate	Water	[134]
Silicone resin containing hydrocarbon groups substituted with amino group or ammonium group.	Insulating liquids	Not mentioned	Water	[135]
Polyhydric alcohols, copolymers polymerized from acrylic acid, methyl acrylate, methyl methacrylate, olefins, vinyl acetate, maleic acid, maleic anhydride, N-vinyl pyrrolidone, methacrylic acid, divinyl benzene, or diallyl ether	All prior art	All prior art	Water, metal ions, organic ions	[136]

Table 2. (continued).

Particle	Fluids	Surfactants	Additives	Ref. #
Polyhydric alcohol, lithium polymethacrylate, mixture of polymethacrylate and methylene bis-acrylamide	Halo and alkyl-substituted diphenyl alkanes, diphenyl ethers, diphenyl sulphones, diphenyl sulfoxides, hydrocarbons, fluorinated polymers (i.e., trifluorovinyl chloride), polychlorinated biphenyls	Not mentioned	Water	[137]
Phenolformaldehyde polymers, dilithium salt of 2,2,4,4-tetra-hydroxy benzophenone and formaldehyde mixture	Diaryl compounds, brominated diphenyl methane	Not mentioned	Water	[138]
Porphin, azaporphin systems, phthalocyanine, metallo-phthalocyanine, poly-(acene-quinone) polymers, polymeric Schiff bases, electronic conductor, semiconductor, unsaturated fused polycyclic system	Halogenated aromatic liquid, chlorinated paraffin oil	Not mentioned	"Substantially" anhydrous	[139]
Silica gel	Silicone oil	Amino/hydroxy/acetoxy or alkoxy functional polysiloxanes	Water	[140]
Silica gel	Liquid hydrocarbons, paraffins, olefins, aromatics	Nitrogen and hydroxyl containing hydrocarbon polymers	Water	[141]
Aluminum silicate (surface Al/Si = 0.15–0.80)	All prior art, silicone oil, isododecane	Hydroxy-, acetoxy-, amino-, and alkoxy-functional polysiloxanes, sorbitan sesquioleate, tetradecylamine, 2-heptadecenyl-4,4(5H)-oxazole dimethanol	Not mentioned	[142]
Hydrophobic polymer particle core (i.e., acrylates, methacrylates, styrenes, vinylacetate, vinyl chlorides, etc.), with a neutralized hydrophilic polymer shell (i.e., acrylic acid, methacrylic acid, aconitic acid, fumaric acid, 2-sulfoethyl acrylate, 2-sulfoethyl methacrylate, p-vinyl benzene sulfonate, etc.)	Insulating medium (i.e., mineral oils, silicone oils, halogenated aromatic liquids, halogenated paraffins and mixtures)	Steric stabilizers, homopolymers or copolymers [i.e., poly(methyl methacrylate), poly(vinyl acetate), poly(12-hydroxystearic acid) and poly(lauryl methacrylate)]	Water, neutralizing agents (i.e., alkyl and aryl compounds of Li, Al, Zn, Mg, B, Na, and K, lithium tertbutoxide, octadecylamine, pyridine, benzyl trimethyl ammonium hydroxide)	[143]
Three component particle: 1. solid organic core: including polystyrene, polyethylene, polyamide, phenol resin, cellulose, starch; 2. conductive film: such as metals, metal compounds, organic conductors; 3. insulating film: made from polyvinyl chloride, polyamide, polyacrylonitrile, polyfluorovinylidene, wax, silica, alumina, titanium dioxide, barium titanate	All prior art, dibutyl sebacate	Not mentioned	"Free" of moisture	[144]

(continued)

Table 2. (continued).

Particle	Fluids	Surfactants	Additives	Ref. #
Metal cation containing zeolite	All prior art	Not mentioned	"Substantially" free of water	[145]
Lithium hydrazinium sulfate	All prior art, silicone oil	Amino-functionalized poly-siloxane, block copolymers	"Absence" of water	[109]
All prior art, starch, silica gel, organic polymers, mono-saccharide, copolymers of phenol and aldehydes	Fluorosilicones, polychlorotri-fluoroethylene	Not mentioned	Not mentioned	[146]
Activated analcime-type zeolite, amorphous silica	Not mentioned	Not mentioned	Not mentioned	[147]
Acid group containing poly-mers, silica gel, starch, elec-tronic conductor	Polyfluoroalkylmethylsilox-anes, such as poly-methyl 3,3,3-trifluoropropylsiloxane, polymethyl-1,1,2,2-tetra-hydroperfluorooctylsiloxane	Not mentioned	Metal cations	[148]
Conductive core of metal/ alloy (i.e., Al, Zn, Fe, W, Pb, Cu-Ni, bronze, steel, monel, etc.), Si, Ge or liquid (i.e., salt water, sulfuric acid, HCl, or acetic acid) with an insu-lating shield (i.e., polyure-thane, nitrile elastomers, epoxies, nylon, ceramics, clay, cement, silica, rubber, teflon, glass, etc.)	Prior art (i.e., silicone oil, paraffin oil)	Incorporation of glass spheres or trapped gas in insulating shield	Not mentioned	[149]
Particles containing electro-lytic solution obtained by hy-drolysis and condensation of metal alkoxides or deriva-tives	Not mentioned	Not mentioned	Not mentioned	[150]
Anionic surfactants (i.e., fatty acid salts, alkyl arylsul-fonates, alkyl sulfates, sul-fonated amides, amines, or esters, sodium dodecyl sul-fate, sodium dodecylbenzene sulfonate, and algenic acid)	Mineral oil, white oil, paraffin oil, chlorinated hydrocar-bons, silicones, transformer oils, halogenated aromatic liquids, halogenated paraf-fins, polyoxyalkylenes fluori-nated hydrocarbons	Steric stabilizer (i.e., amino-, hydroxy-, acetoxy-, and alkoxy-functional polysilox-anes, graft or block copoly-mers	Water, ethylene glycol, or diethylamine	[151]
Metals (i.e., Al, Cr, Ti, Sn), semiconductors (i.e., carbon, molybdenum disulfide, Si, Ge), dichroic crystals of vari-ous geometries (i.e., plates, rods)	Perchloroethylene and ester mixture, silicone oil, ali-phatics, esters, plasticizers, halogenated aliphatics and aromatics, fluorocarbons	Polymeric surfactants	Not mentioned	[152]
Polyhydric alcohol (i.e., monosaccharide), water con-taining polymer particles, Li polymethacrylate (LiPM), LiPM crosslinked with methylene bis-acrylamide	Diaryl derivatives, hydrocar-bons, fluorinated polymers, polychlorinated biphenyls	Not mentioned	Not mentioned	[153]
Carbonaceous particulates and optically anisotropic spherules with insulating layer (i.e., high mw materi-als, silane coupling agents, modified silicone oils, sili-cone surface active agents, inorganic oxides)	Silicone oil	Not mentioned	Not mentioned	[154]

Table 2. (continued).

Particle	Fluids	Surfactants	Additives	Ref. #
Silica gels, hydrous resins, diatomaceous earth, alumina, silica-alumina, zeolites, ion exchange resins, cellulose	Naphthene type mineral oils, paraffin type mineral oils, poly-α-olefins, polyalkylene glycols, silicones, diesters, polyol ester, phosphoric acid esters, silicon compounds, fluorine compounds, halogenated aromatic liquids, polyphenyl ethers, synthetic hydrocarbons	Magnesium sulfonate, calcium sulfonate, calcium phenate, calcium phosphonate, polybutenyl succinic acid imide, sorbitol monooleate, sorbitol sesquioleate	Water, polyhydric alcohols, acids (i.e., sulfuric, HCl, nitric, perchloric, chromic, phosphoric, acetic, formic, etc.), bases (i.e., NaOH, KOH, etc.), salt (i.e., LiCl, KI, K_2SO_4, etc.), metal salts of formic, acetic, oxalic or succinic acid	[155]
Reaction product of a polymeric or monomeric crown ether and a quaternary amine	High dielectric strength, low dielectric constant oils	Not mentioned	Not mentioned	[156]
Solid electroconductive particles (i.e., Al, Si, Ni, Al-Si alloy, carbon black, graphite, SiC, polyacetylene, polypyrrole, copper sulfide, indium oxide, etc.) or particles with conductive coating over a core of synthetic polymers, cellulose, silica, alumina, etc., with an insulating shield (i.e., polyvinyl chloride, polyamide, polyacrylonitrile, polyvinylidene, fluoride, wax, asphalt, varnish, silica, alumina, titanium dioxide, barium titanate, etc.)	Insulating oil (i.e., diphenyl chloride, butyl sebacate, alcohol esters of aromatic polycarboxylic acid, halophenyl alkyl ether, trans-oil, chorinated paraffin, fluoro-oil, silicone oil, etc.).	Surfactants	Rust inhibitors, "virtually free of water"	[157]
Amorphous silicas, synthetic silicas, precipitated silicas, silicates, aluminum silicates, polymethacrylic acid salts, etc., where conductivity is between 10^{-5}–10^{-8} S/m	Mineral oils, white oils, paraffin oils, chlorinated hydrocarbons, silicone oils, transformer oils, halogenated aromatic liquids, halogenated paraffins, polyoxyalkylenes, fluorinated hydrocarbons, etc., where refractive index mismatch with particle is less than 0.02	Surfactants, such as functionalized polydimethylsiloxanes or glycerol esters, etc.	Optional activators, such as water, alcohols, amine compounds, etc.	[158]
Clay, mixed metal hydroxide consisting of $Li_{(0-1)}D_{(0-4)}T(OH)_{(\geq 3)}A_a^n$ and $Mg_{1.7}Al_{0.5}(OH)_{5.0}$, where D = divalent ion of Mg, Ca, Ba, Sr, Mn, Fe, Co, Ni, Cu or Zn a = number of anions A T = metal ion of Al, Ga, Cr or Fe A = ion other than hydroxyl n = valence of A ion	Mineral oil, silicone-based oils, alcohols, polyols, glycols, hydrocarbons, halogenated hydrocarbons, greases, aldehydes and ketones	Aliphatic carboxylic acid	Particle is complexed with an amine salt	[159]
Graphite fibers	Kerosene, benzene, dielectric oil	Not mentioned	Not mentioned	[160]
Aluminum silicate (zeolite) with Si:Al > 8:1	Silicone oil, hydrocarbon oil	Non-ionic surfactants	Water	[161]

(continued)

Table 2. (continued).

Particle	Fluids	Surfactants	Additives	Ref. #
Starch, silica gel, monosaccharide, phenol-formaldehyde copolymer, polymer of acrylate or methacrylate salt	Pentachlorophenyl alkyl ethers, 2,4,6-tribromophenyl alkyl ethers, pentafluorophenyl alkyl ethers, mineral oil, vegetable oil, liquid fluoropolymers, polychlorinated biphenyl, halosubstituted diphenyl methane	Not mentioned	Not mentioned	[162]
60 wt.% hygroscopic inorganic particles, such as crystalline zeolites	40 wt.% electric insulating oils	Not mentioned	10 wt.% water and adsorbed polar compounds, such as ethylene and propylene carbonates, etc.	[163]

zero-field and field dependent properties of an ER material is in the form of a family of shear stress versus shear strain rate curves as shown in Figure 7 for Lord ERF/6184-86B. This type of format also allows for a description of both the dynamic and static yield stress values exhibited by the ER material.

A trade-off between lowering the viscosity of the ER material and limiting the degree of particle settling needs to be recognized. The determination of the rate of particle settling as well as abrasiveness in an ER material is highly dependent upon the application and device design because of the extreme variations in geometries and shear rates encoun-

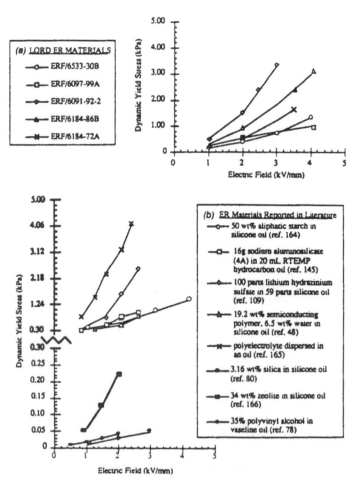

Figure 6. *Comparison of dynamic yield stress as a function of electric field for (a) several ER materials developed by Lord Corporation and (b) ER materials reported in literature.*

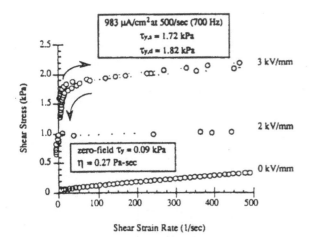

Figure 7. *Shear stress versus shear strain rate data obtained for Lord ERF/6184-86B.*

conductivity of an ER material formulation is important to adequately predict the power consumption in a particular device design. The difference between DC and AC conductivity must be recognized. The measurement of conductivity upon the application of a DC electric field (Curves D–N in Figure 8) represents energy that must be dissipated or removed from the system by some method. However, the conductivity measurement obtained for an ER material subjected to an AC electric field (Curves A–C in Figure 8) does not necessarily represent the amount of energy that will need to be dissipated. With an AC field, the measured current is largely a displacement current, which depends primarily on the capacitance of the device.

tered. Thus, evaluation of these material characteristics must be done in conjunction with the lifetime testing performed on specific device components.

CURRENT DENSITY MEASUREMENTS

Most of the ER material formulations described in the published literature exhibit current densities in the range of 10^{-6}–10^{-3} amp/cm² as shown in Figure 8. A measure of the

RESPONSE TIME ESTIMATES

In order for an ER material to be utilized in continuously variable, controllable devices, a response time on the order of a few milliseconds is required. It has been assumed that the response times associated with known ER materials are within this time limit. The origin of this assumption can be traced back to the results disclosed in the initial work by Winslow [29]. It is often not recognized that response time values [29,35–37,93,169,170] are only applicable to the specific material compositions that were evaluated. A variety of ER materials prepared in this laboratory have demonstrated response times that range from less than a milli-

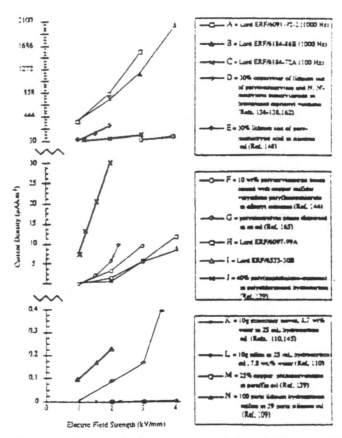

Figure 8. *Comparison of current density versus electric field strength for several ER materials reported in literature and developed by Lord Corporation.*

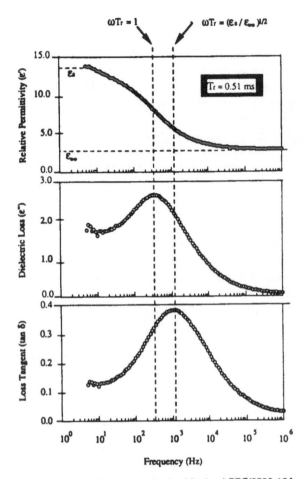

Figure 9. *Dielectric spectra obtained for Lord ERF/6533-16A.*

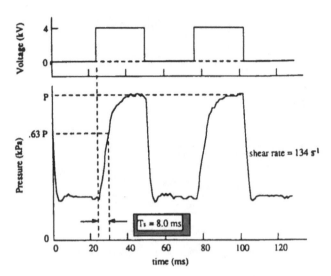

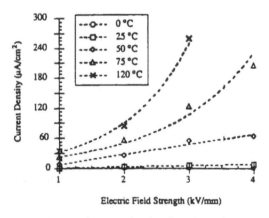

Figure 10. *Response time data obtained on fixed electrode flow fixture for Lord ERF/6533-16A.*

second to greater than one second. An estimate of the relaxation time constant (T_r) associated with the polarization decay process in an ER material can be obtained from the dielectric spectrum exhibited by a particular material as shown in Figure 9. The relaxation time constant obtained from such a spectrum is a measure of the time necessary for individual particles to respond to a small change in electric field. Thus, it provides a lower limit for the time interval in which an ER material response can be expected. Knowledge of this lower limit is advantageous for the design of continuously variable devices.

In devices that are designed for on/off applications, it is essential to know the time required to reach the maximum control (shear stress) ratio. As shown in Figure 10, estimates of the single time constant response (T_s) and the time required for an ER material to reach rheological equilibrium can be obtained by measuring pressure build-up as a function of time for the material flowing through a valve. The overall "turn-on" time of an ER material is generally taken to be 2–3 single time constants. While this single time constant is related to the relaxation time constant obtained from the dielectric data, it includes additional time necessary for macroscopic structure formation in the material, as well as any time required to accommodate mechanical compliance

in the test device. For example, the single time constant for Lord ERF/6533-16A was measured to be 8.0 msec (Figure 10) as compared to the relaxation time constant of 0.5 msec determined from the dielectric spectra (Figure 9). The single time constant arising from the time dependent pressure measurements have been reported to be similar to those obtained in actual devices operated at similar shear rates [35].

TEMPERATURE STABILITY

The various ER materials described in the literature can be divided into two distinct categories: (1) materials that rely on an ionic polarization mechanism; and (2) materials that utilize electronic or nomadic polarization mechanisms. This categorization is important in estimating the temperature limitations expected for different material formulations. The current density associated with a typical ionically polarizable ER material is observed to be highly dependent upon temperature, as shown in Figure 11. It is a serious concern in the design of devices that an ER material be stable over

Figure 11. *Measured current density plotted as a function of temperature for Lord ERF/6097-96B.*

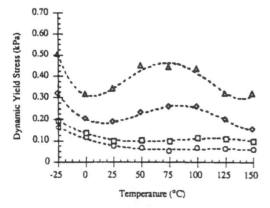

| Percent Change in Current Density (μA/cm²) | | | |
Temperature (°C)	–O– 0.0 kV/mm	–□– 1.0 kV/mm	–◇– 2.0 kV/mm	–△– 3.0 kV/mm
-25	--	-3	-3	-6
0	--	-3	-3	-2
25	--	0	0	0
50	--	+3	+3	+3
75	--	+5	+8	+6
100	--	+4	+13	+6
125	--	+5	+9	+17
150	--	+5	+17	+23

Figure 12. *Dynamic yield stress and current density measurements obtained as a function of temperature for Lord ERF/6533-01A.*

Pre-Yield Properties

In comparison to the previously described post-yield behavior data available for the controllable viscoelastic material properties exhibited by ER materials in the pre-yield regime is severely limited [41–52]. Although the mechanism associated with this pre-yield behavior is not completely understood, it is generally agreed that the complex modulus of ER materials is dependent upon electric field, the strain frequency, and the strain amplitude. It is hoped that work currently in progress will explain the cause/effect relationship for this phenomenon in more detail. Since the yield stress of an ER material becomes larger and the yield strain remains constant upon increasing the strength of the applied electric field, the complex shear modulus (G^*) has also been observed to increase. In fact, the complex shear modulus can vary by several orders of magnitude over an electric field range of 0–4 kV/mm. The pre-yield properties observed for Lord ERF/03-145 are shown in Figure 13 as a function of frequency. The transition of an ER material from viscoelastic behavior to plastic behavior has been reported to occur at strain levels of less than 10% [49,50]. Upon comparing the pre-yield properties of ER materials with the properties exhibited by other viscoelastic damping materials, it is obvious that the storage modulus (G') of the ER materials are many orders of magnitude less than common viscoelastic solids. The continual development of ER materials exhibiting higher yield stress values coupled with the possibility of controlling this property under certain conditions supports further evaluation of these materials in damping applications.

the necessary temperature range. It is unfortunate that very little research describing the dependence of mechanical/electrical properties on operational temperature has been reported. Most of these published results center around an evaluation of ER materials containing alumino-silicate particles [145,171] at elevated temperatures (>100°C). In addition, many formulations have been disclosed that incorporate additives into the material in order to increase the high temperature range of operation. Although these materials operate in a substantially "anhydrous" fashion at elevated temperatures, their utility at low temperatures (<0°C) has not been established.

An ER material that demonstrates the ability to operate over a large temperature range or survive exposure to broad temperature extremes is desirable for many applications. The preparation of this type of ER material is no longer a near term goal, but rather a current accomplishment. ER materials that exhibit the ability to survive prolonged exposure to and operation at broad temperature extremes have recently been identified. For example, Lord's ERF/6184-86B, ERF/6184-72A and ERF/6091-92-2 are capable of being exposed and operated at very low (−25°C) and high (125°C) temperatures without any detrimental effect to their performance at 25°C. Furthermore, it currently is possible to prepare ER materials that can operate with a relatively constant dynamic yield stress and less than a 25% change in current density over a temperature range of −25 to 150°C, as shown in Figure 12 for Lord ERF/6533-01A.

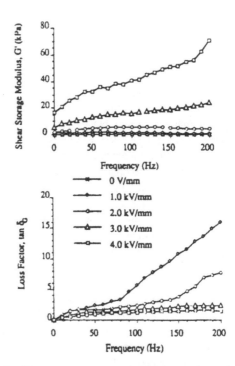

Figure 13. *Measured shear storage modulus and loss factor for Lord ERF/03-145 plotted as a function of frequency.*

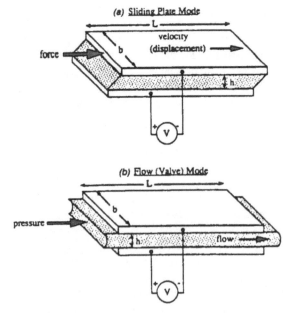

Figure 14. *Schematic illustration of two general electrode configurations used in ER material devices: (a) sliding plate; (b) fixed plate.*

MECHANICAL TEST METHODOLOGY

There are a variety of techniques that can be used to obtain property information about an ER material. Mechanical tests on ER materials may be divided into two broad classes which parallel the two general ways in which ER materials are used. These are described schematically in Figure 14. Sliding plate tests constrain the material between opposing electrode surfaces that can move relative to one another such that the material is subjected to a more or less uniform shear stress. Sliding plate tests may be steady state or dynamic and may be so arranged as to measure either the pre-yield or post-yield regimes. Sliding plate tests include sliding parallel plate methods, couette and modified couette arrangements, and coaxial cylinder tests used in either a rotary or axial mode. In contrast, flow or valve mode tests cause the ER material to flow between a pair of fixed electrodes, usually in a channel flow geometry. Flow mode test devices may be likened to extrusion-type rheometers.

Instrumentation used to obtain the post-yield regime mechanical data, which has been previously published, includes a sliding plate apparatus [125–128,132,139], a channel flow fixture [109], a cylindrical rotation viscometer [129, 133–138,140–142,144,146,162], a Weisenberg rheogoniometer [145] and a modified couette rheometer [111,148]. The methods reported for the measurement of response times associated with ER materials include an optical level torque tube [169], rheo-optical characterization [93], dielectric spectroscopy [35], and channel–flow valve pressure response techniques [35,170]. Finally, several methods also exist through which the pre-yield regime data may be obtained. These techniques include oscillatory flow within a channel [49], shear wave propagation [46], oscillatory par-

allel plate rheometry [42,43,45,48,50,52] and a concentric cylinder annular pumping apparatus [41]. Although each of these methods may have inherent benefits regarding their utilization, further discussion will concentrate on describing in more detail the modified couette rheometer, the channel flow fixture technique, the annular pumping apparatus, and dielectric spectroscopy.

Concentric Cylinder Couette Cell Rheometer

Since the techniques that utilize the cylindrical viscometer, rheogoniometer and the couette rheometer are similar, further description of a commercial concentric cylinder rheometer is warranted. The theory which provides the basis for this technique is adequately described in the literature [172–174]. The information that can be obtained from a concentric cylinder rheometer includes data relating mechanical shear stress to shear strain, the static yield stress, and the electrical current density as a function of shear rate. For ER materials, the shear stress versus shear rate data can be modeled after a Bingham plastic in order to determine the dynamic yield stress and viscosity. Commercial rheometers can be obtained that either operate in controlled shear rate or controlled shear stress modes. A rheometer that can control stress over a wide shear rate range for materials with differing effective viscosities is capable of accurately determining static yield stresses. However, this type of rheometer usually lacks the stability to test materials with very high dynamic yield stresses and low viscosities. Although a rheometer that can control shear rate is incapable of determining static yield stress, its ability to handle higher torque levels and its stability during testing allows accurate determination of large dynamic yield stresses.

The test geometry that is utilized by these rheometers for the characterization of ER materials is a simple concentric cylinder couette cell configuration as shown in Figure 15. The material is placed in the annulus formed between an inner cylinder of radius R_1 and an outer cylinder of radius R_2. One of the cylinders is then rotated with an angular velocity

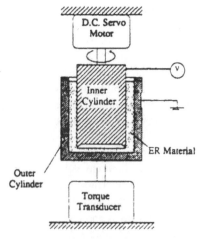

Figure 15. *Schematic of concentric cylinder couette cell.*

Ω while the other cylinder is held motionless. The relationship between the shear stress and the shear strain rate is then derived from this angular velocity and the torque applied to maintain or resist it.

Lord Flow Fixture Apparatus

An apparatus capable of measuring electrical and mechanical properties of ER materials under channel flow conditions has been developed at Lord Corporation [115]. The flow conditions that prevail in such an apparatus provide a good approximation for the flow conditions expected in many "valve" type applications. The test most commonly performed with this type of instrument is the measurement of yield stress and current density as a function of applied voltage at a constant flow rate. The appearance of any particle/oil separation or particle filtering during channel flow can easily be detected during this routine test [115]. This flow fixture apparatus is based upon the establishment of a steady state fluid flow through a channel with a rectangular cross section and the subsequent measurement of the pressure drop along the channel resulting from this flow state. The construction of the instrument, as shown in Figure 16, includes two opposing conductive walls along the length of the channel enclosure in order to apply an electric field to the ER test material.

Assuming Bingham plastic behavior, a relationship between pressure drop and yield stress can be derived for an ER material undergoing channel flow. The details of this derivation were first presented by Phillips [175] in 1969. Phillips' analysis leads to a cubic equation describing the relationship between pressure gradient, p', and yield stress, τ_y. In non-dimensionalized form, this relationship can be written as

$$P^3 - (1 + 3T)P^2 + 4T^3 = 0 \qquad (3)$$

where P and T are a non-dimensionalized pressure gradient and yield stress defined by

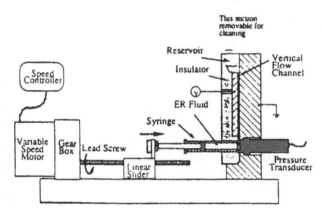

Figure 16. *Schematic of flow fixture apparatus developed by Lord Corporation.*

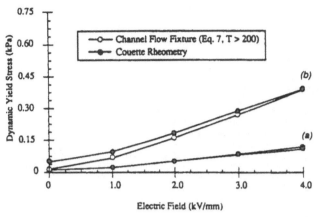

Figure 17. *Comparison of dynamic yield stress for (a) Lord ERF/ 6097-38A and (b) Lord ERF/6097-47B measured using flow fixture apparatus and concentric cylinder rheometry.*

$$P = \frac{bh^3 p'}{12Q\eta} \qquad (4)$$

and

$$T = \frac{bh^2 \tau_y}{12Q\eta} \qquad (5)$$

in which b and h are the flow channel width and height, respectively, and Q is the volume flow rate. Phillips shows that for $T < 0.5$, the relationship between pressure gradient and yield stress may be approximated by the asymptotic solution given in Equation (6). While for $T > 200$, the asymptotic solution given in Equation (7) is a good approximation.

$$T \approx (1/3)P - 1/3 \quad ; \quad T < 0.5 \qquad (6)$$

$$T \approx (1/2)P - 1/3 \quad ; \quad T > 200 \qquad (7)$$

The flow fixture apparatus is designed such that Equation (7) provides a good approximate solution for most ER materials. Dynamic yield stress values obtained with the flow fixture apparatus using this relationship agree very closely with results obtained using a concentric cylinder rheometer, as shown in Figure 17. A more complete derivation of these two relationships [Equations (6) and (7)] is provided in our second paper concerning device construction and potential applications [6].

For testing of an ER material's response time [35], the motor controller for the flow fixture apparatus is set to move the plunger at a linear speed (44 mm/s) that will produce a flow rate necessary to achieve a shear rate of 537 s^{-1} or 134 s^{-1} for 1 mm and 2 mm gap heights, respectively. A 20 Hz, unipolar square wave with a peak voltage of 4 kV is then applied to the flow fixture. This test is appropriate for ER materials that exhibit response times between about 1 ms and 25 ms. For slower ER materials, it is necessary to decrease the frequency of the high voltage signal. In addition, the speed of the plunger must be decreased so that at least several cycles of the high voltage signal can be applied within the allowable travel range of the plunger. For example, a plunger speed of 4 mm/s and a 0.125 Hz unipolar square wave can be used to test materials having a response time on the order of several seconds.

Dielectric Spectroscopy

The measurement of the dielectric spectra for ER materials can be obtained through the use of an impedance analyzer [35]. The impedance parameters that are typically measured include capacitance and conductance. From these parameters, the relative permittivity dielectric loss, loss tangent, and conductivity of the ER material can be calculated. The excitation signal delivered to the sample in these measurements usually has an amplitude of 1.0 volt. In addition, the frequency of the excitation signal is usually stepped from 5 Hz to 1 MHz, with 20 increments per decade. An entire spectrum can be obtained in approximately 2 minutes.

The sample cell designed for the measurement of the dielectric properties of ER materials is shown in Figure 18. This cell consists of a hollow aluminum bob inserted into a cup sized to provide a 0.25 mm gap at the bottom and sides of the bob. The conical shape at the bottom of the bob insures that air bubbles are not trapped. The upper portion of the bob has a reduced diameter so that the measurements are not sensitive to small variations in the actual volume of material placed in the cell. The bob is purposely made hollow in order to reduce the thermal mass of the cell and allow rapid heating and cooling. All aluminum surfaces in contact with the ER material are highly polished and all inside corners are provided with a generous radius to facilitate cleaning between samples. The bob and insulators are connected to the lid by a single, large, knurled nut. The lid and bob assembly is secured to the cup by several thumb screws. The impedance analyzer is connected to the sample cell by a pair of wires that plug into small holes in the bob stem and lid.

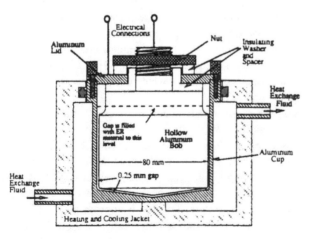

Figure 18. *Schematic of sample cell for impedance measurements with ER materials.*

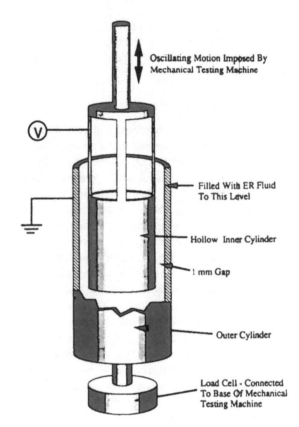

Figure 19. *Schematic of annular pumping apparatus.*

Annular Pumping Apparatus

A concentric cylinder annular pumping apparatus that can be used to determine the complex shear modulus is currently under development by Lord Corporation [41]. This apparatus is schematically depicted in Figure 19. This instrument consists of a hollow aluminum cylinder that is connected to the actuator of a hydraulic mechanical testing machine, and an ER material chamber connected to a load cell having a force range of 0–110 N. An electric field is maintained in the gap between the cylinders with the inner cylinder being displaced vertically using noise with a bandwidth of 0–200 Hz. The complex shear modulus is then deduced from the force and displacement waveforms resulting for each set of test conditions. The results are usually presented in the form of shear storage modulus, G', and the shear loss factor, $\tan \delta_G$, as a function of frequency.

SUMMARY AND CONCLUSIONS

The primary barriers to establishing a commercial ER business have been the lack of satisfactory materials and an inadequate understanding of the ER phenomenon. Although no consensus regarding the mechanism for the observed ER effect has been reached, it is generally accepted that this phenomenon originates from particle polarization induced by an electric field. The observed changes in mechanical properties exhibited by an ER material are a direct result of the formation and breakdown of the induced particle chain network. Current research efforts directed towards providing experimental evidence for the validation of theoretical models will be useful in determining the relationships that exist between individual particle/fluid parameters and the resulting properties exhibited by the formulated ER material.

It is difficult to compare the mechanical and electrical properties that have been reported for different ER materials because of the lack of any uniformity in test methodology and electrorheological response definition. It is suggested that the minimum information needed by an engineer to adequately evaluate a particular ER material within specific device designs includes shear stress versus shear strain rate data, dynamic and static yield stress values (if Bingham plastic behavior is observed), a measure of the zero-field viscosity, an estimate of response time, current density measurements as a function of shear rate, and a measure of the pre-yield complex shear modulus.

Calculations regarding the minimum levels for macroscopic properties needed by ER materials to perform within various device constraints can be found in the literature. ER materials that exhibit greater than 3 kPa in dynamic yield stress, less than 0.35 Pa·s in zero-field viscosity, and stability over a broad temperature range are currently available. ER materials that exhibit these properties will undoubtedly increase the practicality of a variety of existing device designs. It is only a matter of time before the commercialization of ER material technology becomes a reality.

NOMENCLATURE

η = viscosity (Pa·s)

τ = shear stress (kPa)

γ = shear strain

$\dot{\gamma}$ = shear strain rate (sec^{-1})

$\tau_{y,s}$ = static yield stress (kPa)

$\tau_{y,d}$ = dynamic yield stress (kPa)

J = current density (μA/cm^2)

T_s = single time constant response (ms)

T_r = relaxation time (ms)

G^* = complex shear modulus

G' = storage modulus (kPa)

G'' = loss modulus (kPa)

$\tan \delta_G$ = modulus loss factor

ϵ' = relative permittivity

ϵ'' = dielectric loss

$\tan \delta$ = loss tangent

REFERENCES

1. (a) 1988. *Smart Fluids—New Route to Advanced Hydraulic Systems/Devices, Emerging Technologies No. 35*, Technical Insights, Inc. Fort Lee, NJ: Englewood; (b) 1988. *Business Opportunities in Electrorheological Fluids and Devices*. Falls Church, VA: Technology Catalysts, Inc.
2. Scott, D. and J. Yamaguchi. 1985. *Automotive Engineering*, 93(11): 75–90.
3. Jordan, T. C. and M. T. Shaw. 1989. *IEEE Transactions on Electrical Insulation*, 24(5):849–878.
4. Block, H. and J. P. Kelly. 1988. *J. Phys. D: Appl. Phys.*, 21:1661–1677.
5. Gast, A. P. and C. F. Zukoski. 1989. *Advances in Colloid and Interface Science*, 30:153–202.
6. Coulter, J. P., K. D. Weiss and J. D. Carlson. In press. "Engineering Applications of Electrorheological Materials", *Journal of Intelligent Material Systems and Structures*.
7. Duff, A. W. 1896. *Phys. Rev.*, 4:23.
8. Stuetzer, O. M. 1959. *J. Appl. Phys.*, 30(7):984–994.
9. Hogan, J. D. and D. L. Edwards. 1961. *J. Appl. Phys.*, 32(9):1784.
10. (a)Andrade, E. N. da C. and C. Dodd. 1939. *Nature*, 143(3610):26–27; (b) Andrade, E. N. da C. and C. Dodd. 1946. *Proc. Roy. Soc.*, A187:296–337; (c) Andrade, E. N. da C. and C. Dodd. 1951. *Proc. Roy. Soc.*, A204:449–464; (d) Andrade, E. N. da C. and J. Hart. 1954. *Proc. Roy. Soc.*, A225:463–472.
11. (a) Björnstähl, Y. 1935. *Physics*, 6:257–264; (b) Björnstähl, Y. and K. O. Snellman. 1937. *Kolloid.-Z.*, 78:258–272; (c) Björnstähl, Y. and K. O. Snellman. 1939. *Kolloid.-Z.*, 86:223–230.
12. Alcock, E. D. 1936. *Physics*, 7:126–129.
13. Sosinski, S. 1939. *Nature*, 144:117.
14. (a) Honda, T., T. Sasada and T. Sakamoto. 1979. *Jpn. J. Appl. Phys.*, 18(6):1031–1037; (b) Honda, T. and T. Sasada. 1977. *Jpn. J. Appl. Phys.*, 16(10):1775–1783; (c) Atten, P. and T. Honda. 1982. *J. Electrostat.*, 11(3):225–245.
15. Priestley, J. 1769. *The History and Present State of Electricity with Original Experiments, 2nd Ed*. London. [On Readex microprint, 1974, *Landmarks of Science*, New York or London.]
16. Winckler, F. 1748. *Essai sur la Nature, Effets et les Causes de l'Electricite, Vol. 1*. Paris: Sebastian Jorry. [On Readex microprint, 1974, *Landmarks of Science*, New York or London.]
17. Manegold, E. 1950. *Kolloid.-Z.*, 118:11–26.
18. Liebesny, P. 1939. *Arch. Phys. Ther.*, 19:736.
19. Heller, J. H. 1959. *Digest 12th Annual Conference on Electronic Technology in Medicine and Biology* (IRE-AIEE-ISA):56.
20. Teixera-Pinto, A. A., L. L. Nejelski, J. L. Cutler and J. H. Heller. 1960. *Exp. Cell Res.*, 20:548.
21. Soyenkoff, B. C. 1931. *J. Phys. Chem.*, 35:2993–3010.
22. Muth, E. 1927. *Kolloid-Z.*, 41:97–102.
23. Sher, L. D. 1963. "Mechanical Effects of A.C. Fields on Particles Dispersed in a Liquid: Biological Implications", Ph.D. thesis, University of Pennsylvania.
24. Krasny-Ergen, W. 1936. *Hochfreq. Elektoak.*, 48:126.
25. (a) Saito, M. and H. P. Schwan. 1960. In *Biological Effects of Microwave Radiation, Vol. 1*, M. F. Peyton, ed., New York: Plenum Press, p. 85; (b) Schwarz, G., M. Saito and H. P. Schwan. 1965. *J. Chem. Phys.*, 43:3562; (c) Saito, M., H. P. Schwan and G. Schwarz. 1966. *Biophys. J.*, 6:313.
26. Pohl, H. A. 1978. *Dielectrophoresis*. London: Cambridge University Press, p. 495.
27. Winslow, W. M. U.S. patent 2,417,850, 1947.
28. Winslow, W. 1990. In *Proceedings of the 2nd International Conference on ER Fluids*, J. D. Carlson, A. F. Sprecher and H. Conrad, eds., Lancaster, PA: Technomic Publishing Co., Inc., IX–XII.
29. Winslow, W. M. 1949. *J. Appl. Phys.*, 20:1137–1140.
30. Duclos, T. G. 1988. Technical paper 881134, Society for Automotive Engineers, Warrendale, PA, pp. 1–5.
31. Hartnett, J. P. and R. Y. Z. Hu. 1989. *J. Rheology*, 33(4):671–679.
32. Lingard, S., W. A. Bullough and L. S. Ho. 1991. *Wear*, 142:373–381.
33. Hartsock, D. L., R. F. Novak and G. J. Chaundy. 1991. *J. Rheology*, 35(7):1305–1326.

34. Webb, N. 1990. *Chemistry in Britain* (April):338–340.

35. Weiss, K. D. and J. D. Carlson. In press. In *Proceedings 3rd International Conference ER Fluids*, R. Tao, ed., New York: World Scientific Publishing Company.

36. Klass, D. L. and T. W. Martinek. 1967. *J. Appl. Physics*, 38(1):67–74.

37. Klass, D. L. and T. W. Martinek. 1967. *J. Appl. Physics*, 38(1):75–80.

38. Uejima, H. 1972. *Jpn. J. Appl. Physics*, 11:319–326.

39. Hasted, J. B. 1989. *Aqueous Dielectrics*. London: Chapman and Hall.

40. Smyth, C. P. 1955. *Dielectric Behavior and Structure*. New York: McGraw-Hill.

41. Coulter, J. P., T. G. Duclos and D. N. Acker. 1989. Presented at *Damping '89, Palm Beach, FL, 1989.*

42. Gamota, D. R. and F. E. Filisko. 1991. *J. Rheology*, 35(7):1411–1425.

43. Gamota, D. R. and F. E. Filisko. 1991. *J. Rheology*, 35(3):399–425.

44. Korobko, E. V., A. D. Matsepuro and I. P. Krasnikova. 1989. *Vesti Akadmii Navuk BSSR, Ser Fiz-Energ Navuk, Minsk*, 2:66–71.

45. Connolly, M. and R. Marsh. 1991. *American Laboratory—News Edition* (August).

46. Brooks, D., J. Goodwin, C. Hjelm, L. Marshall and C. Zukoski. 1986. *Colloids and Surfaces*, 18:293–312.

47. Shul'man, Z. P., E. V. Korobko and Yu. G. Yanovskii. 1989. *Journal of Non-Newtonian Fluid Mechanics*, 33:181–196.

48. Xu, Y. and R. Liang. 1991. *J. Rheology*, 35(7):1355–1373.

49. Thurston, G. B. and E. B. Gaertner. 1991. *J. Rheology*, 35(7):1327–1343.

50. Yen, W. S. and P. J. Achorn. 1991. *J. Rheology*, 35(7):1375–1384.

51. Vinogradov, G. V., Z. P. Shulman, Yu. P. Yanovski, V. V. Barancheeva, E. V. Korobko and I. V. Bukovich. 1986. *Inzh.-Fiz. Zh.*, 50:605.

52. Chung, K. U.S. patent 4,994,198, 1991.

53. Arguelles, J., H. R. Martin and R. Pick. 1974. *J. Mech. Eng. Sci.*, 16(4):232–238.

54. Klingenberg, D. J. and C. F. Zukoski. 1990. *Langmuir*, 6:15–24.

55. Schwan, H. P. and L. D. Sher. 1969. *J. Electrochem. Soc.*, 116(1):170–174.

56. (a) Adriani, P. M. and A. P. Gast. 1988. *Phys. Fluids*, 31(10):2757–2768; (b) Adriani, P. M. and A. P. Gast. 1989. *J. Chem. Phys.*, 91(10):6282–6289.

57. (a) Klingenberg, D. J., F. van Swol and C. F. Zukoski. 1989. *J. Chem. Physics*, 91(12):7888–7895; (b) Marshall, L., C. F. Zukoski and J. Goodwin. 1989. *J. Chem. Soc. Farad. Trans.*, 85(9):2785–2795; (c) Klingenberg, D. J., F. van Swol and C. F. Zukoski. 1991. *J. Chem. Phys.*, 94(9):6160–6169; (d) Klingenberg, D. J., F. van Swol and C. F. Zukoski. 1991. *J. Chem. Phys.*, 94(9):6170–6178; (e) Klingenberg, D. J., D. Dierking and C. F. Zukoski. 1991. *J. Chem. Soc. Faraday Trans.*, 87(3):425–430.

58. Stanway, R. and J. E. Mottershead. 1989. *J. Modal Analysis* (July):89–92.

59. Eige, J. J. 1964. *Am. Soc. of Mech. Eng.*, Publication 63-MD-1:1–5.

60. (a) Tao, R., J. T. Woestman and N. K. Jaggi. 1989. *App. Phys. Lett.*, 55(18):1844–1846; (b) Jaggi, N. K., J. T. Woestman and R. Tao. 1990. In *Proceedings 2nd International Conference on ER Fluids*, J. D. Carlson, A. F. Sprecher and H. Conrad, eds., Lancaster, PA: Technomic Publishing Co., Inc., pp. 53–62.

61. (a) Jones, T. B. 1990. In *Proceedings 2nd International Conference on ER Fluids*, J. D. Carlson, A. F. Sprecher and H. Conrad, eds., Lancaster, PA: Technomic Publishing Co., Inc., pp. 14–26; (b) Jones, T. B. 1986. *J. Appl. Phys.*, 60(7):2226–2230.

62. Bonnecaze, R. T. and J. F. Brady. 1990. In *Proceedings 2nd International Conference on ER Fluids*, J. D. Carlson, A. F. Sprecher and H. Conrad, eds., Lancaster, PA: Technomic Publishing Co., Inc., pp. 27–40.

63. Wang, K. C., R. McLay and G. F. Carey. 1990. In *Proceedings 2nd International Conference on ER Fluids*, J. D. Carlson, A. F. Sprecher and H. Conrad, eds., Lancaster, PA: Technomic Publishing Co., Inc., pp. 41–52.

64. Korobko, E. V. and Z. P. Shul'man. 1990. In *Proceedings 2nd International Conference on ER Fluids*, J. D. Carlson, A. F. Sprecher and H. Conrad, eds., Lancaster, PA: Technomic Publishing Co., Inc., pp. 3–13.

65. Evans, M. W. and D. M. Heyes. 1991. *J. Phys. Chem.*, 95:5287–5292.

66. Haung, Z. and J. H. Spurk. 1990. *Rheologica Acta*, 29:475–481.

67. Halsey, T. C. and W. Toor. 1990. *J. Statistical Physics*, 61(5/6):1257–1281; (b) Halsey, T. C. and W. Toor. 1990. *Phys. Review Letters*, 65(22):2820–2823.

68. McLeish, T. C. B., T. Jordan and M. T. Shaw. 1991 *J. Rheology*, 35(3):427–448.

69. Whittle, M. 1990. *J. Non-Newtonian Fluid Mechanics*, 37:233–263.

70. Chen, Y., A. F. Sprecher and H. Conrad. 1991. *J. Appl. Physics*, 70(11):6796–6803.

71. (a) Smoluchowski, H. von. 1916. *Kolloid.-Z.*, 18:190–185; (b) Smoluchowski, H. von. 1916. *J. Chem. Soc.*, 110(2):473–474

72. Kransy-Ergen, W. 1936. *Kolloid.-Z.*, 74:147–172.

73. Booth, F. 1950. *Proc. Roy. Soc. London*, A203:533–551.

74. (a) Harmsen, G. J., J. van Schooten and J. Th. G. Overbeek. 1953. *J. Colloid Sci.*, 8:64–71; (b) Harmsen, G. J., J. van Schooten and J. Th. G. Overbeek. 1953. *J. Colloid Sci.*, 8:72–79.

75. Takashima, S. and H. P. Schwan. 1985. *J. Biophysical Soc.*, 47:513.

76. Zimmerman, U. 1983. *Trends in Biotechnology*, 1:149

77. Foster, W. C. 1948. *Technical News Bull. J. Research Bur. Stand.*, 32(May):54–60.

78. (a) Deinega, Yu. F. and G. V. Vinagradov. 1984. *Rheo. Acta*, 23:636–651; (b) Deinega, Yu. F. and T. A. Zharinova. 1978. *Int. Congr. Surf.—Act. Subst.*, 7th Moscow, 2:527; (c) Deinega, Yu. F. and N. Ya. Kovganich. 1975. *Proc. Int. Conf. Coll. Surf. Sci.*, 1:42; (d) Deinega, Yu. F., K. K. Popko and N. Y. Kovganich. 1978. *Heat Transfer-Sov. Res.*, 10:50.

79. Stangroom, J. E. 1983. *Phys. Technol.*, 14:290–296.

80. (a) Sprecher, A. F., J. D. Carlson and H. Conrad. 1987. *Mater. Sci. Eng.*, 95:187–197; (b) Conrad, H., M. Fisher and A. F. Sprecher. 1990. In *Proceedings 2nd International Conference on ER Fluids*, J. D. Carlson, A. F. Sprecher and H. Conrad, eds., Lancaster, PA: Technomic Publishing Co., Inc., pp. 63–81.

81. Cerda, C. M., R. T. Foister and S. G. Mason. 1981. *J. Coll. Int. Sci.*, 82(2):577–579.

82. Shul'man, Z. P., A. D. Matsepuro, L. N. Novichenok, S. A. Demchuk and I. L. Svirnovskaya. 1974. *J. Eng. Phys.*, 27(6):1569–1572.

83. Smith, K. L. and G. G. Fuller. 1987. In *1st International Conference ER Fluids*. Raleigh, NC: NCSU Engineering Publications, pp. 27–46.

84. Korobko, E. V. and I. A. Chernobai. 1985. *J. Eng. Phys.*, 48(2):153–157.

85. (a) Bezruk, V. I., A. N. Lazerev, V. A. Malov and O. G. Usyarov. 1972. *Coll. J.*, 34:142–146; (b) Bezruk, V. I., A. N. Lazarev, V. A. Malov and O. G. Usyarov. 1972. *Coll. J.*, 34:276–280

86. (a) Vorob'eva, T. A. and I. N. Vlodevets. 1974. *Kolloidn. Zh.*, 36(6):1154–1156; (b) Vorob'eva, T. A. and I. N. Vlodevets. 1969. *Kolloid. Zh.*, 31(5):668–673.

87. Gamayunov, N. I. and V. A. Murtsovkin. 1982. *J. Eng. Phys.*, 43(3):963–965.

88. Krasikov, N. N. and A. E. Kovylov. 1970. *J. Appl. Chem. USSR*, 43:1851.

89. Sasada, T., T. Kishi and K. Kamijo. 1974. *Proc. 17th Japan Cong. Mater. Res.*, pp. 228–231.

90. Martin, H. R. and G. Zanetel. 1975. *Am. Soc. of Mech. Engineers*, Publication 75-DE-62:1–8.

91. Gleb, V. K. and S. A. Demchuk. 1972. *Inzh.—Fiz. Zh.*, 23(4):681–685.

92. Sugimoto, N. 1977. *Bull. JSME*, 20(149):1476–1483.

93. Jordan, T. C. and M. T. Shaw. 1990. In *Proceedings 2nd International Conference on ER Fluids*, J. D. Carlson, A. F. Sprecher and H. Conrad, eds., Lancaster, PA: Technomic Publishing Co., Inc., pp. 231–251.

94. Hill, J. C. and T. H. van Steenkiste. 1991. *J. Appl. Phys.*, 70(3):1207–1211.

95. (a) Pohl, H. A. 1958. *J. Appl. Physics*, 29(8):1182–1188; (b) Pohl, H. A. and J. P. Schwar. 1959. *J. Appl. Phys.*, 30:69; (c) Pohl, H. A. 1961. *J. Appl. Phys.*, 32(9):1784–1785.

96. Tao, R. and J. M. Sun. 1991. *Phys. Re Letters*, 67(3):398–401

97. Shul'man, Z. P., V. I. Kordonskii, E. A. Zaltsgendler, I. V. Prokhorov, B. M. Khusid and S. A. Demchuk. 1986. *Int. J. Multiphase Flow*, 12(6):935.

98. Schwarz, G. 1962. *J. Phys. Chem.*, 66 2636–2642.

99. Hanai, T. 1961. *Bull. Inst. Chem. Kyoto Univ.*, 39:341.

100. Dukhin, S. S., T. S. Sorokina and T. L. Chelidze. 1969. *Kolloid. Zh.*, 31(6):823–830.

101. Bickerman, J. J. 1940. *Trans-Farad Soc.*, 36:154.

102. Deryagin, B. V. and S. S. Dukhin. 1969. *Kolloid. Zh.*, 31(3):350–358.

103. Petrzhik, G. G., O. A. Chertkova and A. A. Trapeznikov. 1982. *Coll. J.*, 44:68–74.

104. VanBeek, L. K. H. 1967. *Prog. Dielect.*, 7:69.

105. (a) Dukhin, S. S. 1973. *Surf. Colloid Sci*, 3 83; (b) Deryagin, B V., S. S. Dukhin. 1974 *Surf. Colloid Sci* 7:322; (c) Deryagin, B. V., S. S. Dukhin and V. N. Shilov 1980 *Adv. Colloid Interface Sci.*, 13:141.

106. (a) Stangroom, J. E. 1984. In *Proceedings Conf. Mat. Eng.* London: Inst. of Metallurgists, pp. 81–85; (b) Stangroom, J. E. Publ. Brochure–Ministry of Defence, Agreement #AT/2031 094 RAR: 1 11.

107. Wong, W. and M. T. Shaw. 1990. In *Proceedings 2nd International Conference on ER Fluids*, J. D. Carlson, A. F. Sprecher and H. Conrad, eds., Lancaster, PA: Technomic Publishing Co., Inc., pp. 191–195.

108. Block, H. and J. P. Kelly. 1989. In *Proceedings 1st International Symposium on ER Fluids*, H. Conrad, A. F. Sprecher and J. D. Carlson, eds., Raleigh, NC: NCSU Engineering Publications, pp. 1–26.

109. Carlson, D. U.S. patent 4,772,407, 1988.

110. Filiski, F. E. and L. H. Radzilowski. 1990. *J. Rheology*, 34(4):539–552.

111. (a) Block, H., J. P. Kelly, A. Qin and T. Watson. 1990. *Langmuir*, 6:6–14; (b) Block, H. and J. P. Kelly. 1985. *Proc. IEE Colloq. on Electrically Active Fluids*, 14:1–3.

112. Pohl, H. A. 1978. *Dielectrophoresis*. London: Cambridge University Press.

113. (a) Okagawa, A. and S. G. Mason. 1974 *J. Coll. Interface Sci.*, 47(2):568–587; (b) Arp, P. A. and S. G. Mason 1977. *Coll. Polym. Sci.*, 255(6):566–584; (c) Arp, P. A., R T. Foister and S G. Mason. 1980. *Adv. Coll. Int. Sci.*, 12(4):295; (d) Arp, P. A. and S G. Mason 1977. *Coll. Polym. Sci.*, 255(12) 1165 1173

114. (a) Block, H., W. D. Ions, G. Powell, R. P. Singh and S M Walker. 1976. *Proc. R. Soc. London Ser.* 4, 352.153–167, (b) Block, H., E. M. Gregson, A. Ritchie and S. M. Walker. 1983 *Polymer*, 24:859–864; (c) Block, H., W. D. Ions and S. M Walker. 1978. *J. Polym. Sci., Polym. Phys Ed.*, 16:989 998; (d) Block, H. 1979. *Adv. Polymer Sci.*, 33:93–167, (e) Block, H., E. M. Gregson, A. Qin, G. Tsangaris and S. M. Walker. 1983. *J. Physics E*, 16(9):896–902; (f) Block, H., E. M. Gregson and S M Walker. 1978. *Nature*, 275:632–634.

115. Bares, J. E. and J D. Carlson 1990. In *Proceedings 2nd International Conference on ER Fluids*, J D. Carlson, A F. Sprecher and H. Conrad, eds., Lancaster PA Technomic Publishing Co, Inc, pp. 93–114.

116. Stangroom, J E. 1990. In *Proceedings 2nd International Conference on ER Fluids*, J D. Carlson, A F. Sprecher and H. Conrad, eds., Lancaster, PA: Technomic Publishing Co., Inc, pp 199 206.

117. Seed, M., G. S. Hobson and R C. Tozer. 1990. In *Proceedings 2nd International Conference on ER Fluids*, J. D. Carlson, A F. Sprecher and H. Conrad, eds, Lancaster, PA Technomic Publishing Co., Inc., pp. 214–230.

118. Kraynik, A. M. 1990. In *Proceedings 2nd International Conference on ER Fluids*, J. D. Carlson, A F. Sprecher and H Conrad, eds., Lancaster, PA: Technomic Publishing Co., Inc., pp. 445–454.

119. Zukoski, C. F. and J. W. Goodwin. 1986. *IECON*, pp. 9–13.

120. Anon. 1988 *Engineering* (London), 228(2):i–iv.

121. Winslow, W. U.S. patent 2,611,596, 1953.

122. Winslow, W. U.S. patent 2,611,825, 1953.

123. Winslow, W. U.S. patent 3,047,507, 1962.

124. (a) Klass, D. and T. Martinek. U.S. patent 3,250,726, 1966; (b) Klass, D. and T. Martinek. G.B. patent 1,076,754, 1967; (c) Martinek, T., D. Klass and H. Folkins. U.S. patent 3,412,031, 1968.

125. Martinek, T., D. Klass and H. Folkins. U.S. patent 3,367,872, 1968.

126. Klass, D. and V. Brozowski. U.S. patent 3,385,793, 1968.

127. Martinek, T. U.S. patent 3,397,147, 1968.

128. Martinek, T. and D. Klass. U.S. patent 3,399,145, 1968.

129. Peck, R. U.S. patent 3,427,247, 1969.

130. Clark, H. U.S. patent 3,484,162, 1969.

131. (a) Carreira, L. and V. Mihajlov. G.B. patent 1,178,301, 1970; (b) Carreira, L. and V. Mihajlov. U.S. patent 3,553,708, 1971.

132. Westhaver, J. U.S. patent 3,970,573, 1976.

133. Takeo, K. and Y. Omura. U.S. patent 3,984,339, 1976.

134. (a) Stangroom, J. U.S. patent 4,033,892, 1977; (b) Stangroom, J. G.B. patent 1,501,635, 1978; (c) Stangroom, J. DE patent 2,530,694, 1976.

135. (a) Ishino, Y., S. Endo, T. Osaki, H. Nomura, T. Chikaraish and S. Tomita. G.B. patent 2,217,344, 1989; (b) Ishino, Y., S. Endo, T. Osaki, H. Nomura, T. Chikaraish and S. Tomita. DE patent 3,912,888, 1989.

136. (a) Stangroom, J. U.S. patent 4,129,513, 1978; (b) Stangroom, J. G.B. patent 1,570,234, 1980.

137. (a) Stangroom, J. G.B. patent 2,100,740, 1983; (b) Stangroom, J. U.S. patent 4,502,973, 1985.

138. (a) Stangroom, J. G.B. patent 2,119,392, 1983; (b) Harness, I. and J. Stangroom. U.S. patent 4,483,788, 1984.

139. (a) Block, H. and J. Kelly. G.B. patent 2,170,510, 1986. (b) Block, H. and J. Kelly. E.P.C. patent 0,191,585, 1986; (c) Block, H. and J. Kelly. U.S. patent 4,687,589, 1987.

140. (a) Goossens, J. E.P. patent 170,939, 1986; (b) Goossens, J., G. Oppermann, W. Grape and V. Hartel. U.S. patent 4,645,614, 1987.

141. (a) Goossens, J. E.P. patent 201,827, 1986; (b) Goossens, J., G. Oppermann, W. Podszun and V. Hartel. U.S. patent 4,668,417, 1987.

142. (a) Goossens, J., G. Oppermann and W. Grape. DE patent 3,536,934, 1987; (b) Goossens, J., G. Oppermann and W. Grape. U.S. patent 4,702,855, 1987.

143. (a) Ahmed, S. M. U.S. patent 4,992,192, 1991; (b) Ahmed, S. M. U.S. patent 5,073,282, 1991.

144. Inoue, A. JP patent 63,097,694, 1988.

145. (a) Filisko, F. and W. Armstrong. U.S. patent 4,744,914, 1988; (b) Filisko, F. and W. Armstrong. U.S. patent 4,879,056, 1989; (c) Filisko, F. and W. Armstrong. EP patent 265,252, 1988; (d) Filisko, F. and W. Armstrong. EP patent 313,351, 1989.

146. (a) Stangroom, J. EPC patent 0,284,268, 1988; (b) Stangroom, J. U.S. patent 4,812,251, 1989.

147. (a) Suzuki, T. FR patent 2,612,910, 1988; (b) Nakazawa, T., M. Ogawa, K. Abe, K. Suzuki and J. Suzuki. DE patent 3,811,136, 1988.

148. Brooks, D. and S. Kandian. EPC patent 0,311,984, 1988.

149. Reitz, R. P. WO90/00583, 1990.

150. Hattori, E. M. K. and Y. M. K. Oguri. EP patent 341,737, 1989.

151. Carlson, J D. U.S. patent 5,032,307, 1991.

152. Marks, A. U.S. patent 4,442,019, 1984.

153. Stangroom, J. WO 82/04442, 1982.

154. Ishino, Y., T. Osaki, S. Endo, S. Tomita, T. Maruyama, Y. Fukuyama and T. Saito. EP patent 0,361,106, 1990.

155. Kanbara, M., N. Yoshimura, J. Mitsui and H. Hirano. EP patent 0,342,041, 1989.

156. Cipriano, R. A. U.S. patent 5,071,581, 1991.

157. Prendergast, M. EP patent 0,396,237, 1990.

158. Carlson, J. D. and J. E. Bares. U.S. patent 5,075,021, 1991.

159. Knobel, T. M. and R. A. Cipriano. U.S. patent 5,032,308, 1991.

160. Pedersen, N. U.S. patent 4,737,886, 1988.

161. (a) Gillies, D., L. Satcliffe and P. Bailey. G.B. patent 2,219,598, 1989; (b) Gillies, D., L. Satcliffe and P. Bailey. EP patent 350,167, 1990; (c) Gillies, D., L. Satcliffe and P. Bailey. AU patent 8,936,309, 1989.

162. (a) Stangroom, J. G.B. patent 2,153,372, 1985; (b) Stangroom, J. and I. Harness. EPC patent 0,150,994, 1985.

163. Fukuyama, Y., Y. Ishino, T. Osaki, T. Maruyama and T. Saito. U.S. patent 5,075,023, 1991.

164. Stevens, N. G., J. L. Sproston and R. Stanway. 1987. *Journal of Applied Mechanics*, 54:456–458.

165. Korane, K. J. 1991. *Machine Design* (May):52–57.

166. Conrad, H., Y. Chen and A. F. Sprecher. 1990. In *Proceedings 2nd International Conference on ER Fluids*, J. D. Carlson, A. F. Sprecher and H. Conrad, eds., Lancaster, PA: Technomic Publishing Co., Inc., pp. 252–264.

167. Gow, C. J. and C. F. Zukoski. 1990. *Journal of Colloid and Interface Science*, 136(1):175–188.

168. Treasurer, U. Y., F. E. Filisko and L. H. Radzilowski. 1991. *J. Rheology*, 35(6):1051–1069.

169. Stangroom, J. E. 1989. In *Proceedings First International Symposium on Electrorheological Fluids*, H. Conrad, A. F. Sprecher and J. D. Carlson, eds., Raleigh, NC: NCSU Engineering Publications, pp. 81–98.

170. Peel, D. J. and W. A. Bullough. 1990. In *Proceedings of the Second International Conference on ER Fluids*, J. D. Carlson, A. F. Sprecher and H. Conrad, Lancaster, PA: Technomic Publishing Co., Inc., pp. 141–157.

171. Conrad, H., A. F.Sprecher, Y. Choi and Y. Chen. 1991. *J. Rheology*, 35(7):1393–1410.

172. Oka, S. 1960. In *Rheology, Theory and Applications, Volume 3*, F. R. Eirich, ed., New York: Academic Press.

173. Middleman, S. 1968. *The Flow of High Polymers: Continuum and Molecular Rheology*. New York: Interscience Publishers.

174. Barnes, H. A., J. F. Hutton and K. Walters. 1989. *An Introduction to Rheology*. New York: Elsevier Science Publishers.

175. Phillips, R. W. 1969. "Engineering Applications of Fluids with a Variable Yield Stress", Ph.D. dissertation, University of California, Berkeley.

Effect of Flow Rate, Excitation Level and Solids Content on the Time Response in an Electro-Rheological Valve

D. J. PEEL AND W. A. BULLOUGH

Department of Mechanical and Process Engineering
The University of Sheffield
P.O. Box 600
Sheffield, S1 4DU, United Kingdom

ABSTRACT: Diverse regimes of the fast time response of an electro-rheological fluid are identified in a presentation of experimental results which show the effects of the application of step and DC biased sine wave excitations to a series of set flows in one valve. The form of pressure response is complex and depends to some extent on the rate of flow and applied electric field magnitude. Some frequency domain behaviour is related to the step wave performance. Apart from their value as a foundation study for a new area of rheology, the results are important to the rapidly developing subject of flexibly operated smart machines and as an aid in target setting to developers of the hydraulic semi-conductors on which some of them are based.

INTRODUCTION

THE search for flexibly operated machines which have a better performance than those based on electro-magnetic transmission in terms of speed, torque/inertia ratio, and controllability of force and/or motion has identified the need for some kind of hydraulic semi-conductor. At the present stage of development, this is exemplified by the electro-rheological fluid (ERF) (Bullough, 1990).

Some progress has been made towards understanding the rheology of these fluids in general (Misc., 1989, 1991), the fundamentals of their engineering application (Bullough and Peel, 1990) and the relationship their performance bears to controlling excitation, rate of movement, and the material constituents of the solid/liquid mixtures. This has mainly been in the steady state regime and has been directed at raising the level of the voltage controlled yield stress.

Whilst this is necessary from the point of view of developing the high specific traction forces necessary to accelerate or decelerate any associated controlled element in quick time (t^{**}), the high speed machines requirement has opened up a further avenue of investigation. Not only is the rheological flow pattern changed by the excitation applied, but the governing equations must be solved in the fast time domain. Of particular interest is the electron-hydraulic time constant (t^*)—the time lapse between the application of voltage and the occurrence of the steady level of enhanced shear (yield) stress (Bullough et al., 1991).

This latter parameter is explored within the present work, but in respect of only one of the classical modes of E.R. control: the flow mode where fluid flows through near stationary parallel plate electrodes, in this case cylindrical channels with the bounding surfaces electrified.

Results were taken under conditions of strict temperature control and by the utilisation of high natural frequency pressure transducers. Normally, a problem arises from the necessity to dump the charge quickly on switch-off—some kind of drive-down circuit is required if, say, a sharp-edged, square voltage wave is to be achieved. This was avoided in the first of the present tests by the selection of a relatively low current demand fluid of 20% volume/solids concentration.

A specific aim of the tests was to discover the shape of the ER response of the fluid to excitation form and flow condition, and its dependence on the magnitudes of both. At the present juncture, the results relate to several constant speeds of operation, i.e., in the flow rheometer, the rate of flow is fixed and substantially maintained throughout the application of the excitation. For these situations, the output will involve two or three separate response mechanisms and possibly a time lag. Some similar results, from preliminary tests on both Couette and Poiseuille-type apparatus, which show some of the slower time variation problems are given by Peel et al. (1990).

THE TEST ARRANGEMENT

The fluid used throughout was a 20% (by volume or weight) density matched (at 30°C) slurry/colloidal dispersion comprising moist lithium polymethacrylate particles of mean size 7×10^{-6} m carried in a mixture of Fluorolube and dielectric liquid. The excitations were (a) the application of step voltages between zero and a maximum set value and (b) a sinusoidal voltage imposed on a steady DC bias potential.

A mechanically stiff test rig of copper and steel construction was laid out as in Figure 1. Two quartz pressure transducers of approximately 100 KHz natural frequency were connected to the valve inlet and outlet by nylon tubes, the length of which had been adjusted to have negligible distortion effect on the pressures measured. Amplification was included in the signal conditioning unit, but no filtration was involved there or in the instrumentation circuit. All tests

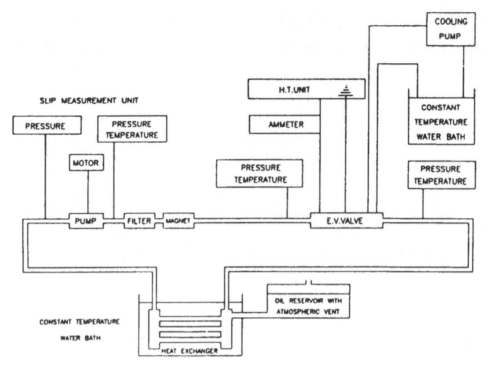

Figure 1. *General layout of valve test rig.*

were run at 30 ± 0.5°C in order to eliminate any severe effects of change in fluid conductance.

Once a potential had been applied, the reading period was
terminated only at a time when the transient had settled. Simultaneous recordings of pressure and voltage with respect
to time were made of the full event and stored on an infinite
persistence oscilloscope. The significant portions of these
were selected and a Polaroid photograph was taken of the

screen display. In every case, the pressure signal at the valve
outlet was observed but, since its fluctuation was always
found to be insignificant, the results were not usually recorded.

A valve of two parallel concentric channels (Figure 2) was
chosen in order to match the load to the driving gear pump.
In this way, the flow could be fixed at a high enough rate to
make the "slip" back past the gears (or more correctly, the

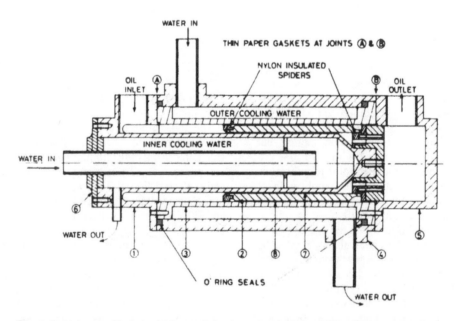

Figure 2. *Valve used in tests: (1) inner electrode and end cap (earthed); (2) central (HT) electrode; (3) outer electrode (earthed); (4) water jacket; (5) end cap; (6) inner water channel end
plate; (7) annular oil channel, inner; (8) annular oil channel, outer.*

change in pump internal leakage with voltage initiated pressure) a very small percentage of the total. To alleviate the change in flow rate with pressure, the pump's DC drive motor was high inertia and used a feedback of speed for correction via thyristor control. Any problems with regard to slip would be more pronounced at the lower speeds (Peel et al., 1991).

The valve channels were of 0.5 mm separation and 100 mm long. High potential was applied to the intermediate electrode, which was 51 mm in diameter inside and 64 mm in diameter outside. Flow rates were chosen to be commensurate with the slip problem and typical engineering scale: 1 to 14 litres/minute. These were set by means of the motor speed controller. The electrical supply system is described in Firoozian et al. (1989).

THE PRESSURE RESPONSE

Within the overall test programme, the responses of the pressure drop across the valve to two types of valve potential signal were investigated. The signal types were (1) a step change of valve electric field/potential gradient E, nominally between $E = 0$ and some maximum magnitude $E = E_{max}$, and (2) a harmonically oscillating electric field/valve potential gradient nominally between $E = 0$ and some maximum magnitude $E = E_{max}$. The same fluid and apparatus was used for both sets of tests.

The peak to peak amplitude of electric field change in either case will be expressed as ΔE, where $E = V/h$ (the difference in interelectrode voltage per unit electrode separation). ΔE is the change in V divided by h. Two nominal ranges of ΔE were investigated: (a) a "low" range where $\Delta E < 2000$ volts/mm, and (b) a "high" range, where $\Delta E < 4000$ volts/mm. A potential gradient of 4000 volts/mm gives the safe pressure limit for the test rig.

The maximum valve potential gradient E, is, in all these tests, negative relative to the zero (earth $E = 0$). This fact is of no significance and results from the earthing requirements of the high tension signal generating equipment. In referring to changes of potential gradient, "increase" will be used to describe an elevation in magnitude (only) from zero of the potential, since the sense of the potential is always negative. Similarly, "decrease" will be used to describe a reduction of magnitude of potential (usually to the nominal zero minimum magnitude).

In the figures relating to the test, usually the only pressure signal shown is that of the valve entrance – the magnitude is shown relative to its zero-volts level. The pressure (absolute) at the valve outlet may be taken in all cases to be nominally atmospheric pressure, within the limits of accuracy of the pressure measurements. The effects of step or harmonic changes of valve potential gradient on the pressure at the valve outlet were negligible, as were the effect of changes of volume flow rate. The data in this paper is extracted from Peel and Bullough (1978).

LOW STEP FIELD CHANGES

The application of a valve potential as a step change from zero to maximum always results in a comparatively slow development of the pressure drop through the valve. Such a

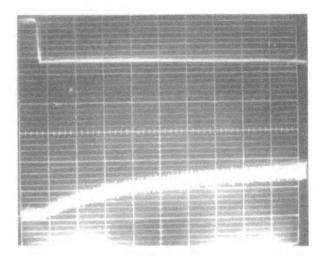

Figure 3. Valve entrance pressure response to potential step increase: Q – 0.9 L/min, ΔE – 600 volts/mm, time scale = 5.0 s/cm, pressure scale – 0.5 bar/cm.

pressure development at the valve entrance is shown in Figure 3 for a step increase $\Delta E = 600$ volts/mm at volume flow rate of $Q \cong 0.9$ L min. The development is still in progress at the end of a period involving tens of seconds. At higher valve potential gradient, large fluctuations were observed in the valve pressure drop, and this phenomenon is illustrated in Figure 4, where the step increase is $\Delta E = 1800$ volts mm at volume flow rate $Q \cong 3.8$ L/min.

The development of the valve pressure drop is of asymptotic form and relatively slow, the initial development being essentially linear with respect to time. This initial "linear" response at the low volume rate $Q \cong 0.9$ L min for potential step increase $\Delta E = 600$ and $\Delta E = 1200$ volts mm, respectively, gives pressure gradients at the valve entrance as 0.04 bar s and 0.3 bar s. At volume flow rate $Q \cong 3.8$ L min for potential gradient step increases, $\Delta E = 600$ and $\Delta E = 1200$ volts mm, respectively, gives pressure gradients at the valve entrance to be 0.07 bar s and 0.7 bar/s.

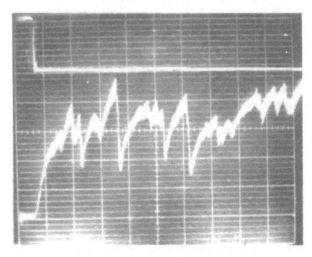

Figure 4. Valve entrance pressure response to potential step increase Q – 3 8 L/min, ΔE 1800 volts/mm, time scale = 5.0 s/cm, pressure scale - 2 0 bar/cm

Thus, in these cases, we can make an approximation: An increase of volume flow rate ×4 has resulted in an increase of the valve entrance pressure time derivative ×2, for a given valve potential gradient, and an increase of valve potential ×2 has resulted in an increase of the valve entrance pressure time derivative ×2³, for a given volume flow rate.

For a valve electric field step increase, $\Delta E = 1200$ volts/mm at volume flow rates $Q \cong 0.9$ and $Q \cong 3.8$ L/min, respectively, there is a small step increase of pressure at the valve entrance. The respective magnitudes of the step are $\cong 0.2$ bar and 0.1 bar.

Responses of the valve entrance pressure to a step decrease in valve potential are inverted. The decay of pressure at the valve entrance is exponential in form, and it appears from inspection of the corresponding developments and decays of pressure that the time for decay of pressure back to the zero volts level is commensurate with the time over which the pressure increased with the voltage applied (when the development time is very small compared with the time required for the pressure to attain its maximum for that volume flow rate and potential). Thus, it appears that the pressure decay rate and total time vary with valve potential gradient and volume flow rate, as does the initial pressure time derivative.

A further point, concerning the step decrease of valve potential, is that there appears to be a short (relaxation) delay when the potential is switched off before the pressure decay begins. In Figure 5, for $Q \cong 3.8$ L/min and $\Delta E = 600$ volts/mm, this delay is $\cong 1.0$ sec., while for $\Delta E = 1200$ volts/mm for the same volume flow rate, the delay is $\cong 0.1$ sec. For lower volume flow rates, any delay is obscured by signal interference.

Further speculation may be made over the valve inlet pressure development at the low volume flow rate $Q \cong 0.9$ L/min at the potential gradient $\Delta E = 1800$ volts/min (not shown). Comparing this with pressure development at $\Delta E = 600$ volts/mm (Figure 3), shows that at high voltage, pressure it is not only irregular, but appears also to be truncated at the point where the irregularity begins. The mean level of the irregular pressure is substantially less than that which would result from extrapolation of the asymptotic form of development occurring at lower voltages. The point of speculation here is whether or not the irregularities of pressure are associated with the diminution of the steady flow valve pressure drop at low flow, and the corresponding diminution of the calculated "yield stress" at the valve plates.

The final point arising from the figures is that there is present interference of a significant level. However, inspection shows that the lower frequency component of the interference corresponds with the rotational speed of the gear pump, and presumably results from eccentricity in the gears, while the higher frequency component of the interference is a multiple of the lower frequency component corresponding with the number of teeth on either pump gear. The peak to peak amplitude of pressure fluctuations at the valve inlet due to eccentricity of the pump gears at $Q \cong 0.9$ L/min is $\cong 0.07$ bar; the peak to peak amplitude of fluctuations due to the passage of the gear teeth is rather more ($\cong 0.1$ bar). The pressure fluctuations will be accompanied

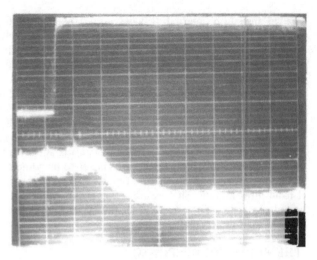

Figure 5. *Valve entrance pressure response to potential step decrease: Q = 3.8 L/min, ΔE = 600 volts/mm, time scale = 0.5 s/cm, pressure scale = 0.2 bar/cm.*

by small fluctuations in the volume flow rate, and there must be some possibility that the irregular valve entry pressure shown in Figure 4 is triggered by the gear pump disturbances. Again, this is a very speculative suggestion, and even if it is the case, there must be some instability in the electro-rheological effect.

HIGH STEP FIELD PRESSURE CHANGES

The response of the valve inlet pressure to step changes of potential of large amplitude, at any given volume flow rate, is composed of at least two parts. Hence, the pressure response for four amplitudes of electric field step changes, $\Delta E = 3600, 2600, 1600,$ and 600 volts/mm at each of flow rates 0.9, 1.9, 7.8, 8.7, and 13.7 L/min, was investigated. A good illustration of the response is given by Figure 6, which shows $\Delta E = 3600$ volts/mm at $Q = 1.9$ L/min. The two parts of the response are as follows: when the potential step

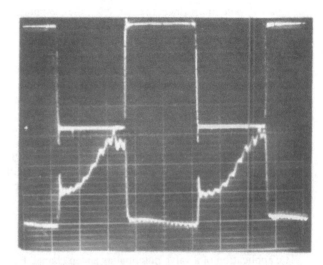

Figure 6. *Valve entrance pressure response to potential step increase and decrease: Q = 1.9 L/min, ΔE = 3600 volts/mm, time scale = 0.2 s/cm, pressure scale = 2.0 bar/cm.*

is increasing, the increase of potential is accompanied by a "step" increase of pressure at the valve inlet, followed by a further increase of pressure, which is roughly linear with time for a short period following the potential gradient change. When the potential step is decreasing, a step decrease of pressure at the valve inlet accompanies the decrease of potential gradient, and a further decay of pressure follows, which is exponential in form but is more or less complete in a time equal to that for which the pressure development lasted.

The step increase and decrease of pressure may be accompanied by a small "overshoot" (Figure 6). This may be associated with overshoot of the field supply signal and/or with wave transmission through the fluid, with inertia, and establishment of the electro-rheological effect. The overshoot is not taken into account in any measurements of pressure change. The step pressure decreases vary in amplitude, but for the larger potential gradient, steps appear to be somewhat greater than the step pressure increase preceding them (i.e., on the incoming wave).

The range of steps in the electric field chosen extends into the "low potential gradient" and at $\Delta E = 600$ volts/mm, the step changes of pressure are absent; only the steady pressure rise component of response follows this small step increase of potential.

The best linearity of the steady pressure rise component of the pressure responses—which will be referred to as the "ramp" pressure rise—occurs for the smaller amplitudes of field step. At larger potential gradient amplitudes, less regularity in the ramp pressure rise may be associated with the irregularity of pressure already noted (in particular for low flow) in Figure 4, for example. In some cases, although there may be a region of quite good linearity of the ramp pressure rise, it is apparently preceded by a short delay in the pressure rise.

The pressure responses at higher flow rates, in particular for lower potential amplitudes, show appreciable gear pump interference. The gear pump interference does not appear to have any effect on the pressure response to the potential changes, however.

In Figures 7 and 8, plots are made respectively of the vari-

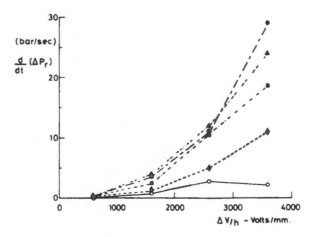

Figure 8. Variation of the initial "ramp" pressure rise derivative at the valve entrance with volume flow rate and amplitude of potential step applied. ○—0.9, ◆—1.9, ■—3.8, ▲—8.7, ●—13.7.

ation, with volume flow rate and amplitude of potential gradient step increase, of the amplitude of the step pressure rise and the mean gradient with respect to time of the ramp pressure rise. The graphs show that in general, the step pressure rise decreases and the ramp pressure rise rate increases as the volume flow rate increases, and the step pressure rise and the ramp pressure rise rate both increase as the potential step amplitude increases.

In general, the influence of the electric field step amplitude is much greater than the influence of volume flow rate. The step pressure rise varies roughly linearly with potential gradient amplitude, being zero for lower field strength. The ramp pressure rise rate varies less than the cube of the potential gradient (which was the variation at low potential) and at higher potentials, the ramp pressure rise rate tends to fall off at low volume flow rates (apparently). Apart from the possible association of the irregular pressure response shown in Figure 4, there may be a contribution to this effect from the fact that as the pump delivery pressure rises, at low flow rates the "slip" flow in the pump rises as a proportion of the flow delivered.

For volume flow rate $Q = 3.8$ L/min and potential step change $\Delta E = 3600$ volts/mm, where the ramp pressure rise rate is $\cong 19$ bar/s (from Figure 8), a comparison can be made of the step pressure rise rate, which is illustrated in Figure 9, with the time scale "stretched" to show the step response in more detail. The rise time for the step potential change is $\cong 6.5$ msec (a mean rise rate of 2.8×10^5 volts/sec for the actual potential step of 1800 volts, the valve gap h being 0.5 mm), while the rise time for the step pressure change of 2.5 bars (including overshoot) is 8.5 msec, giving a mean step pressure rise rate of $\cong 3000$ bar/s—some 160 times the subsequent ramp pressure rise (mean) rate. Figure 9 shows a delay of about 5 msec between the potential and pressure step rise beginnings.

Figure 10 shows the rate of pressure decay on switch-off following the pressure rise on switch-on (shown in Figure 9) to be about 4000 bar/s, while the potential decay period is some 10 ms. A plate taken at the same time shows the pressure response at the valve outlet, corresponding to the valve inlet response at Figure 10, to be virtually zero.

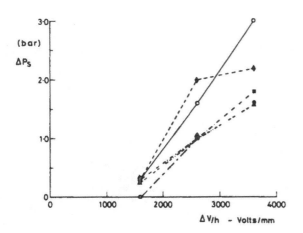

Figure 7. Variation of the "step" pressure rise component at the valve entrance with volume flow rate (L/min) and amplitude of potential step applied. —0.9, ◆—1.9, ■—3.8, ▲—8.7, ●—13.7.

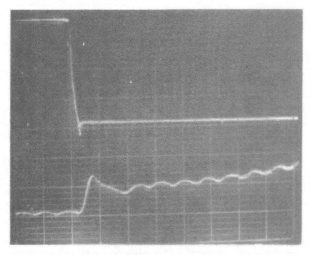

Figure 9. *Valve entrance pressure response to potential step increase:* Q = 3.8 L/min, ΔE = 3600 volts/mm, time scale = 0.02 s/cm, pressure scale = 2.0 bar/cm.

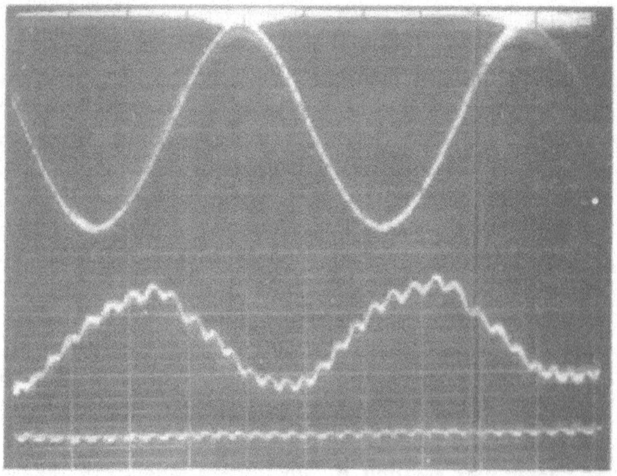

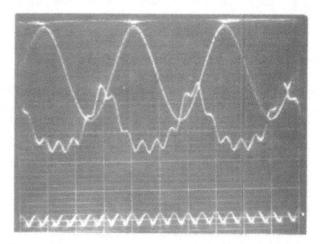

Figure 11. *Valve entrance pressure response to a potential harmonic signal: frequency* f = 1.0 Hz, Q = 0.9 L/min, ΔE = 3600 volts/mm, time scale = 0.2 s/cm, pressure scale = 2.0 bar/cm.

THE HIGH FIELD HARMONIC CHANGES

The responses of electroviscous valve inlet pressure to harmonic changes of potential of large amplitude has been investigated for a single amplitude of the potential gradient, at various frequencies of oscillation, and for six volume flow rates: 0.9, 1.9, 3.8, 6.2, 8.7 and 13.7 L/min. The range of frequencies of the DC biased voltage sine wave was $0.2 \leq f \leq 40$ Hz, and the amplitude of the sine wave potential gradient was $\Delta E = 3600$ volts/mm. The unsteady potential and pressure signals are typically illustrated in Figures 11–14; the sine wave was biased at the $E = 1800$ volts/mm mean level.

The form of the pressure response to significant harmonic potential signals can be seen to be broadly simply harmonic. There was some distortion and erratic pressure change at the lowest frequency 0.2 Hz throughout the range of flow rates. The erratic behaviour is predominantly in the region

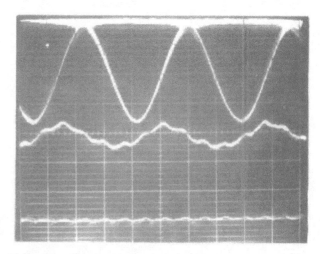

Figure 12. *Valve entrance pressure response to a potential harmonic signal: frequency* f = 15 0 Hz, Q = 6.2 L/min, ΔE = 3600 volts/mm, time scale = 0.02 s/cm, pressure scale = 1.0 bar/cm.

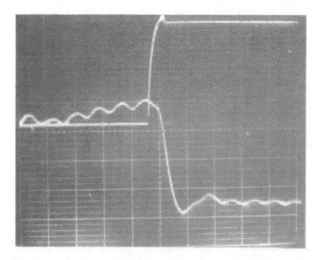

Figure 10. *Valve entrance pressure response to potential step decrease:* Q = 3.8 L/min, ΔE = 3600 volts/mm, time scale = 0.02 s/cm, pressure scale = 2.0 bar/cm.

Figure 13. *Valve entrance pressure response to a potential harmonic signal: frequency* f = 25 0 Hz, Q = 8.7 L/min, ΔE = 3600 volts/mm, time scale = 0.01 s/cm, pressure scale = 2.0 bar/cm.

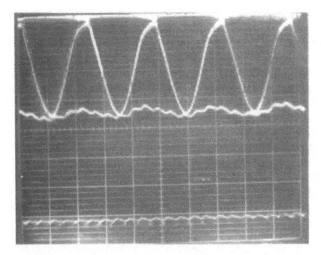

Figure 14. *Valve entrance pressure response to a potential harmonic signal: frequency f = 40.0 Hz, Q = 13.7 L/min, ΔE = 3600 volts/mm, time scale = 0.01 s/cm, pressure scale = 2.0 bar/cm.*

of large but diminishing pressure, where the potential is still of large magnitude, and is perhaps associated with the behaviour illustrated in Figure 4. Further disturbances of the pressure signal are due to the interference of the gear pump, which becomes of significant amplitude and frequency when the potential oscillation frequency is high. There is also some distortion in the potential gradient signal at high frequency; the reason for this is not known, but should not be attributed too readily to any deficiency of the amplifier. Although the specified frequency limit of the amplifier is <200 Hz, it does not seem to have caused any difficulties with the step potential changes. Figures 11 to 14 also show the data (zero volts) levels of both the potential gradient (upper) and pressure (lower).

At higher frequencies of the potential, the oscillating pressure component becomes superimposed on a steady pressure component of large amplitude; this steady component develops in the manner illustrated in Figure 3. The levels of this steady pressure component are of no significance, and appear completely arbitrarily or from a root mean square tendency.

Figure 15 shows the variation, with frequency, for each volume flow rate, of the peak to peak amplitude of the unsteady component of the pressure. This illustration has a number of prominent features: (a) over the lower half of the test frequency range, as the frequency increases, the pressure amplitude steadily diminishes for all volume flow rates; the peak to peak amplitude of the unsteady pressure will be denoted ΔP_u; although only the valve entrance pressure is measured, the result indicates the unsteady pressure loss component through the electroviscous valve; (b) at these lower frequencies, ΔP_u is greater at higher flow rates than at lower volume flow rates; (c) there is for all volume flow rates a sharp leveling off of ΔP_u between 1 and 3 Hz of potential gradient frequency; relatively, this decay of pressure response is sharper for larger volume flow rates; (d) over the upper half of the test frequency range, as the frequency increases, there is a gradual but small decrease of ΔP_u; (e) at high frequency, ΔP_u is commonly greater at lower volume flow rates.

A restriction was placed on the frequencies used because of the capacity of the supply. This tended to be inadequate above 200 Hz, at which point pressure responses tended to be at a low (but practical) level.

LOW FIELD HARMONIC CHANGES

A ΔE = 1200 volts/mm peak to peak input at 1 Hz resulted only in a very low amplitude pressure wave (less than 1 bar p–p) over the whole flow range. In view of the step wave results time derivatives, this is not surprising.

DISCUSSION AND ANALYSIS

General

In the first part of this paper, the use of step voltage inputs to an electro-rheological valve showed severe non-linearities in the pressure response. From another viewpoint, the results gave a valuable insight into the digital latching or pulse width control of electro-rheological devices, which has been pursued elsewhere (Bullough et al., 1991). Ramp voltage inputs have been used to successfully avoid problems involving time dependence when characterising electro-rheological fluids at low voltage/field strengths. Such inputs result in quasi-steady response and by nature do not give any indication of the time response properties of the controller/fluid combination. In using biased sine wave excitation, an attempt has been made to avoid some of the disadvantages of these techniques while producing some detailed quantification of the time response problem.

A normal control system approach would involve using a series of small amplitude harmonic inputs of differing frequencies and bias voltage. From the results of linear analyses across the full bias range, a better picture could be built

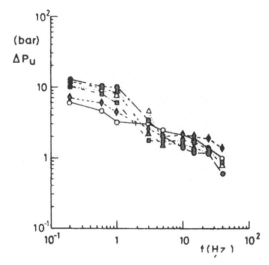

Figure 15. *Variation of the peak to peak unsteady pressure amplitude ΔP_u at the valve entrance with volume flow rate and frequency f of simple harmonic potential applied. Peak to peak potential amplitude: ΔE = 3600 volts/mm. ○—0.9 L/min, ♦—1.9 L/min, ■—3.8 L/min, □—6.2 L/min, ▲—8.7 L/min, ●—13.7 L/min.*

of the true time response properties. This was not possible in the present work, as the level of noise (mainly from the pump) required the sine waves of voltage to be the same order of magnitude as the maximum safe voltage that could be applied. Also, the non-linearity is very pronounced in the region of the threshold or low voltage region where little yield/shear (particularly step response) effect is generated by the field. Thus, comparable conditions (flow rate and field strengths) consistent with those in the step wave tests were used for the sine wave excitation tests. The effects of excitation were extracted from what proved to be substantial harmonic outputs. These are reported as an intermediate stage prior to the testing of a superior fluid.

Step Potential

No estimate has been made of the influence of the complex mechanics of the valve test rig components of the valve inlet pressure response. However, it seems likely that the step pressure rise rate will be slightly damped. This might occur, for example, because the pump seal housing is ventilated and communicates with the valve supply via the pump bearings, or because of elasticity in the system. Further, more damping may occur in the ramp pressure rise rate, since it occupies a more extensive period of time when inertial effects for the whole system and increases of gear pump slip, for example, may be of more significance; the gear motor's speed control system adds further complication in this case. However, there seems to be no evidence of any resonance of any components, or of any gross damping such as might result from the presence of significant amounts of air within the fluid. Therefore, these response tests may be expected to give a good idea of the orders of magnitude of pressure changes and rate of change associated with step changes of valve potential.

Several phases of response are shown in the flow mode tests described above. The apparent time lag between the voltage trigger and the onset of pressure change needs further exploration and quantification yet, in terms of high speed machine application, the time for step pressure development t^* is short compared with the (L/R) inductive response time constant of an electro-magnetic event of equal significance. The developed pressure (the specific force accelerating term which would be applied to a piston, for example) determines t^{**}. In turn, this seems to involve at least two separate events with respect to time: a step and a ramp response. Conditions which lead to the step being the greater part of the pressure rise need detailing. This should occur from both the materials point of view and the percentage concentration of solid, etc. It is interesting that Block and Inoue (Misc., 1989, 1991) both postulate that the rate of the charging event may be important, yet in a similar but more dense fluid, Hosseini-Sianaki et al. (1991) witnessed little effect of practical rate of rise times. However, the 30% volume fluid did give a much greater proportion of step (to total) pressure rise under equivalent operating conditions and on the same 100 mm long valve.

On the occurrence of unstable flow (e.g., Figure 4), some elementary flow visualisation was attempted. A perspex tube fitted downstream of the valve showed significant changes in the amount of light transmitted through the fluid.

A valve depressurisation brought with it a distinctly darker shade. From the point of view of machine reliability, some quantification of this critical (choking) event is called for; related phenomena have been noted before (Dienega and Vinogradov, 1984). The shape of the pressure response in all tests suggests that the fidelity of the transducer arrangement was adequate for the purpose of registering this event.

All applications of voltage were in the same direction—there was no alternating current. Thus, the significance of any possible dielectro-phoresis with respect to the slow final phase of pressure buildup cannot be ascertained. The only indication of its absence is the similarity of the response in both the loading and unloading cases.

Harmonic Potentials

In many respects, the results achieved can reasonably be expected. For a given valve potential, it has already been found that the steady flow valve pressure loss and its rate of development are diminished at low volume flow rate for a given valve potential. Further, the maximum period of the unsteady potential in the test range is 5 s, which is sometimes small compared with the valve pressure drop development times illustrated in the step wave tests. These two factors may account for the behaviour at low frequency. However, the diminution of pressure generation between 1 and 3 Hz for all flow rates cannot immediately and obviously be explained in terms of what has been discussed so far, and neither can the behaviour at high frequency.

One difficulty in discussing the pressure response in Figure 15 is that the amplitudes of pressure at any flow rate and frequency are absolute measurements, and are not related to any expectation which might normally be determined as the frequency $f \to 0$. Without some form of "datum" for the pressure amplitude, it is difficult to put the results obtained in context. However, the determination of the same would have exceeded the test rig pressure limit.

There is, however, another way of dealing with this problem. If it is assumed, as it has been observed, that the pressure response has a simple harmonic form, then the unsteady component of pressure at the valve entrance is described by $p = p_{max} \sin 2\pi ft$, where p and t measure pressure and time, respectively and p_{max} is the (half) amplitude of the unsteady pressure component. The time derivative of the pressure is $dp/dt = 2\pi f p_{max} \cos 2\pi ft$.

If the pressure response is dominated by the "ramp" pressure rise (linear portion in step tests), then the gradient of this ramp will represent the maximum expectation of the pressure time derivative for the prevailing electric field amplitude and volume flow rate. This maximum expectation, which will be denoted $(dp/dt)_{max}$, may be taken directly from Figure 8, from which may be found a maximum expectation of the unsteady pressure (half) amplitude, expressed $(dp/dt)_{max}/2\pi f$, or, where $T = 1/f$ is the period of the unsteady potential and $\Delta p_{max} = 2p_{max}$ is the peak to peak amplitude of maximum expectation of the unsteady pressure, $\Delta p_{max} = T(dp/dt)_{max}/\pi$. It will be assumed that the valve entrance pressure response in the test frequency range is dominated by the initial linear portion of the ramp pressure rise component.

The ramp pressure rise gradient has been found to depend

on the potential gradient applied to the valve, and on the volume flow rate. In the present frequency response test, the potential gradient amplitude is constant, so that only the volume flow rate will influence the ramp gradient with consequence. Thus, the significance of the volume flow rate is characterised by the mean "residence" time of particles flowing through the valve. This mean residence time will be written as ΔT, and is determined as

$$\Delta T = \frac{\text{Fluid volume between valve plates}}{\text{Volume flow rate}} = \frac{B\ell h}{Q}$$

where b, ℓ, and h are the valve plate width, length, and separation gap. Note that ΔT does not signify change in T.

The valve entrance pressure response may then be expected to depend on the relationship between the mean fluid residence time in the valve and the period of the potential oscillation. Figure 16 shows a plot of $\Delta P_u/(T/\pi)(dp/dt)_{max}$ versus $\Delta T/T$. It is observed that in general, the data seems to collapse quite well onto a single curve, which suggests that the relationship includes all pertinent factors in two dimensionless ratios. The one exception to this is the case of very low flow rates; the derivation of this result might be explained in part by pump slip (Peel et al., 1991), this not having been taken into account and being quite significant at the pressures occurring in these tests, at low flow rate. A further factor in this low flow case may be the relatively low value of the "ramp" pressure rise component; this will lead to higher values of the ratio $\Delta P_u/(T/\pi)(dp/dt)_{max}$. It could have its origin in the effects of the development of static shear stress or pseudo-hysteresis.

Some of the characteristics of Figure 16 are obscured by the linear scale of $\Delta P_u/(T/\pi)(dp/dt)_{max}$; however, a much more informative plot is shown in Figure 17, where a logarithmic scale is used. The characteristics of the relationship shown there are as follows: (a) at low val-

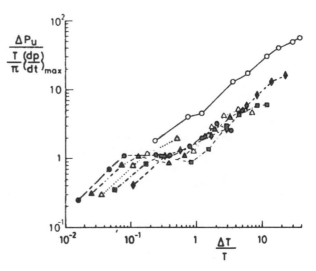

Figure 17. *Variation of dimensionless peak to peak unsteady pressure amplitude $\Delta P_u/(T/\pi)(dp/dt)_{max}$ at the valve entrance with volume flow rate and dimensionless frequency $\Delta T/T$ of simple harmonic potential applied. Peak to peak amplitude of potential $\Delta E = 3600$ volts/mm. —0.9 L/min, ◆—1.9 L/min, ■—3.8 L/min, —6.2 L/min, ▲—8.7 L/min, ●—13.7 L/min.*

ues of $\Delta T/T$, as $\Delta T/T$ increases, $\Delta P_u(T/\pi)(dp/dt)_{max}$ increases until it reaches its maximum of unity; until this point is reached, $\Delta P_u(T/\pi)(dp/dt)_{max}$ is larger for higher flow rates; (b) for a given volume flow rate, when $\Delta P_u/(T/\pi)(dp/dt)_{max}$ has reached unity, as $\Delta T/T$ increases further, $\Delta P_u(T/\pi)(dp/dt)_{max}$ remains stationary at unity until circa $\Delta T/T \cong 0.5$; (c) for $\Delta T/T > 0.5$, as $\Delta T/T$ increases, $\Delta P_u/(T/\pi)(dp/dt)_{max}$ increases steadily above unity and its magnitude does not appear to be too strongly dependent on volume flow rate.

These characteristics might be interpreted as follows:

(a) Where the mean fluid residence time in the valve is very small compared with the period of the oscillating potential, the pressure response at the valve entrance is dominated by the asymptotic or exponential characteristics of development of the valve pressure drop in steady flow following a change of potential. As $\Delta T/T$ diminishes, the initial "ramp" pressure rise gradient represents a rate of pressure development increasingly greater than that which actually occurs, and therefore the ratio $\Delta P_u/(T/\pi)(dp/dt)_{max}$ is diminished. The increase of $\Delta P_u/(T/\pi)(dp/dt)_{max}$ with volume flow rate in this region is assumed to be associated with the characteristics of steady flow, i.e., to be partly due to the effects of "plug" flow in the valve, and perhaps also to the apparent "truncation" of pressure development at low volume flow rates.

(b) Where the mean fluid residence time in the valve approached half the period of the oscillating potential, the pressure response at the valve entrance was dominated by the initial portion of the "ramp" pressure rise component in the development of the valve pressure drop in steady flow following a step change of potential gradient. This ramp pressure rise component represents the maximum attainable rate of development of large

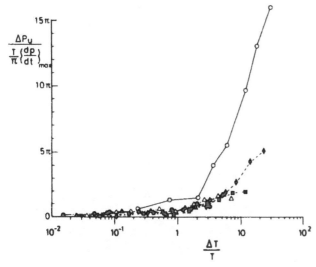

Figure 16. *Variation of dimensionless peak to peak unsteady pressure amplitude $\Delta P_u/(T/\pi)(dp/dt)_{max}$ at the valve entrance with volume flow rate and dimensionless frequency $\Delta T/T$ of simple harmonic potential applied. Peak to peak amplitude of potential $\Delta E = 3600$ volts/mm. —0.9 L/min, ◆—1.9 L/min, ■—3.8 L/min, △—6.2 L/min, ▲—8.7 L/min, ●—13.7 L/min.*

pressure amplitudes. As the period of oscillations diminishes, so does the attainable pressure amplitude, and as $\Delta T/T$ approaches a value of $\cong 0.5$: this attainable pressure amplitude is seen to be reached, but not exceeded (there is one exceptional data point at volume flow rate $Q = 6.2$ L/min, for which there is no account, and as it has been pointed out, there must be substantial error in the low flow rate data—at $Q = 0.9$ L/min—and which therefore cannot be included in this discussion). This explains the observation in Figure 15 that there is an apparent change point in the pressure response development at a frequency of between 1 and 3 Hz.

(c) Where the mean fluid residence time in the valve exceeds half the period of the oscillating potential, the pressure response at the valve entrance is dominated by the "step" pressure rise component in the development of the valve pressure drop in steady flow following a step change of potential gradient. When the residence time of fluid in the valve is large compared with the period of the oscillating potential, there will be no time for any significant ramp type pressure development. But as this component of the unsteady pressure diminishes, it will become less significant than that associated with the step response, whose time of development has been seen to be only a few milliseconds. As frequency increases, then, a roughly constant amplitude of pressure response may be expected until the period of the oscillating potential becomes small relative to the rise time of the pressure step. This roughly constant amplitude is seen at higher frequencies in Figure 15 for all flow rates, and in Figure 17 allows the steady increase of $\Delta P_v/(T/\pi)(dp/dt)$ above unity for $\Delta T/T > 0.5$ (approximately). It was observed in Figure 15 that in this frequency region, the pressure amplitude is greater at lower than at higher volume flow rates. This is a reasonable expectation since in Figure 7, it can be observed that for a given potential amplitude, the pressure step is larger at lower flow rates.

CONCLUSIONS

Analysis of the valve entrance pressure response to unsteady potentials varying in both step and harmonic forms has shown three distinct types of response to harmonic excitation, each associated with a different frequency band. It is clear that fluid specification and valve design in any application would have to be matched carefully to the needs of the application to ensure easily predictable behaviour without the complications which might arise in crossing between the frequency bands. This in general ought not to prove too difficult. Further, it seems that useful high frequency response is being achieved; even though the pressure amplitudes here are small, they are significant and not strongly frequency dependent; this is a useful feature. The upper frequency limit of response has not been determined, and rise times of potential gradient and pressure steps of the same order of magnitude have been observed; more detailed attention to equipment and a test rig/procedure redesign seems necessary before any upper frequency limit is investigated.

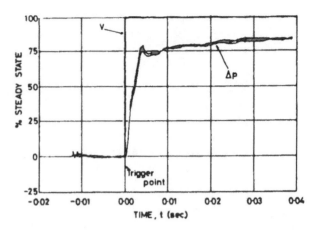

Figure 18. *30% fluid, typical result. Note enhanced step part of response to sudden application of high amplitude electric field gradient at high flow rate. Fluid composition is the same as 20% save for higher solids content.*

One reported test, on a 30% volume fraction but otherwise like fluid, is shown in Figure 18. Although the test rig is similar and the valve is the same as in the present work, the percentage of the response attributable to the fast part of the rise is substantially increased (i.e., beyond that of the 20% fraction fluid). This 30% fluid, which is more promising in terms of t^* and t^{**}, is further investigated in Johnson et al. (1991) and Whittle et al. (1992) for phase lags and non-Newtonian behaviour.

In a quantitative sense, there is little against which to compare the results of the present work. However, there is much more interest in ERF time response characteristics than hitherto before. Very recently, more general works (Misc., 1989, 1991), which use visualisation of structure, have measured stress or calculation of polarisation time responses to give qualified, if loose, support to the above. One very significant point to emerge from these comparisons is the severe effects of scaling. For the time being, if useful engineering data is to be obtained, it is necessary to experiment close to the regime of practical application. Low volume fraction fluids at low excitation potentials may or may not be useful in developing the fundamental understanding of ER fluids operating under engineering conditions.

REFERENCES

Bullough, W. A. 1990. "Liquid State Force and Displacement Devices", *Mechatronics,* 1(1) 1 10.

Bullough, W. A. and D. J. Peel. 1990. "The Field Controlled Liquid State". *Proc. I Mech E., Mechatronics Conf., Cambridge,* pp. 171–177; Firoozian, R., D. J. Peel and W. A. Bullough "Magnetic Effects in an Electro-Rheological Controller." Ibid., pp. 231–238.

Bullough, W. A., R. Firoozian, A. Hosseini-Sianaki, A. R. Johnson, J. Makin and Shi Xiao 1991. "The Electro-Rheological Catch Latch Clutch", *Proc. I Mech E. Eurotech Direct Conference, Birmingham,* pp. 129 136.

Deinega, Y. F. and V. G. Vinogradov 1984 "Electric Fields in the Rheology of Disperse Systems", *Rheol. Acta,* 23:636.

Firoozian, R., D. J. Peel and W. A. Bullough 1989 "Time Domain Modelling of the Response of an Electro-Rheological Fluid in the Flow Mode", in *Vibration Analysis Techniques and Applications DE, Vol. 18 J ASME* pp. 45 50.

Hosseini-Sianaki, A., W. A. Bullough, J. Makin, R. Firoozian and R. Tozer. 1991. "Experimental Measurements of the Dynamic Response of an ERF in the Shear Mode", *Proc. Int. Conf. ER Fluids*. S. Ill. Univ.: World Publ. Co., pp. 219–238.

Johnson, A. R., W. A. Bullough, R. Firoozian, A. H. Hosseini-Sianaki, J. Makin and S. Xiao. 1991. "Electro-Rheological Clutch under Inertial Loading", *Proc. JSME Motion and Power Trans. Conf. Hiroshima.*, pp. 1016–1021; 1991. "Operational Considerations in the Use of an Electro-Rheological Catch Device", *Proc. 1st Fluid Power Trans. and Control Symp, Beijing*, pp. 591–595.

Miscellaneous. 1991. *Proc. Int. Conf. Electro-Rheological Fluids*. S. Ill. Univ.: World Sci. Publ. Co.; 1989. *Proc. 2nd Intl. Conf. Electro-Rheological Fluids, N.C. State Univ.* Lancaster, PA: Technomic Publishing Co., Inc.

Peel, D. J. and W. A. Bullough. 1978. "Unsteady Effects and Choking in the E.V. Flow Mode", University of Sheffield, Department of Mechanical Engineering Research Report DR96A.

Peel, D. J., D. Brooks and W. A. Bullough. 1991. "Experiences in the Pumping of ER Fluid", *Proc. 8th Intl. Conf. Fluid Machinery*. Budapest: Hungarian Aca. Sciences, pp. 362–369.

Peel, D. J., E. V. Korobko and W. A. Bullough. 1990. "The Time Response Sequence of an Electro-Rheological Fluid", *Proceedings of Actuator 90 Congress*, pp. 144–149, VDI/E (German Assoc. Engrs.), Bremen.

Whittle, M., W. A. Bullough, R. Firoozian and D. J. Peel. "Underlying Responses in ER Valve Flow", *Jnl. Intel. Material Systems and Structures* (to appear).

Whittle, M., R. Firoozian, D. J. Peel and W. A. Bullough. 1992. "A Model for the Electrical Characteristics of an ER Valve", *Proc. Int. Conf. ER Fluids*. S. Ill. Univ.: World Sci. Publ. Co., pp. 343–366.

Engineering Applications of Electrorheological Materials

JOHN P. COULTER
Department of Mechanical Engineering and Mechanics
Lehigh University
Bethlehem, PA 18015

KEITH D. WEISS AND J. DAVID CARLSON
Thomas Lord Research Center
Lord Corporation
Cary, NC 27511

ABSTRACT: This article provides a summary of the current state of electrorheological (ER) material applications research and development. The use of ER materials in a variety of intelligent material systems is covered. A description of basic material configurations used in controllable system components, as well as a discussion of controllable devices such as antivibration mounts, clutches, and dampers, is presented. More recent developments in the area of ER material adaptive structures are then reviewed. Concepts underlying such structures are presented, and models that have been developed to simulate the response of such structures are summarized. Throughout the article, an attempt is made to identify current and future key areas of research and development in ER material applications technology.

INTRODUCTION

ELECTRORHEOLOGICAL (ER) materials, which exhibit changes in rheological behavior in the presence of an electric field, have been studied sporadically since Winslow's initial observations in the 1940s [1]. Research and development activities related to these materials and their applications have increased dramatically during the past decade. Proposed ER mechanisms, ER material formulations, observed material properties, and characterization techniques are discussed in a separate work by the present authors [2]. This companion article focuses on the utilization of the controllable behavior exhibited by ER materials in engineering applications.

If commercialized, ER materials will have a significant impact on hydraulic equipment, and will be utilized in the automotive, marine, aerospace, robotics, and machinery industries. Several recent attempts have been made to predict the annual market for ER material based systems. The estimated 1995 U.S. and worldwide market figures for such devices have been stated as $3.5 billion and $20 billion, respectively [3,4]. With only three years remaining, it is unlikely that acceptable devices will be developed in time to capitalize on this suggested 1995 market.

APPLICATION CLASSIFICATION

The controllable rheological behavior of electrorheological materials is useful in engineering systems and structures where variable performance is desired. When a tunable product based on electrorheological behavior is implemented so as to have sensory and control capabilities, the result is an intelligent material system or an adaptive structure. In general terms, such systems or structures are those that can sense external stimuli and react in an appropriate manner so as to optimally meet pre-specified performance criteria.

The response of an ER material to an applied electric field can be observed in a variety of material properties. Changes in the optical, electrical, volumetric, acoustic, and mechanical properties of ER materials have all been observed to date, and applications based on each of these changes have been proposed. Nearly all applications that have been studied in some detail, however, are based on the controllable mechanical response of ER materials. Thus, the primary focus of the present review is ER material controllable mechanical systems.

There are two fundamental ways to effect a change in a mechanical system. One involves the addition of mechanical energy to the system (active control), while the other involves the modification of the mechanical properties of the system (semi-active control). Since piezoelectric, electrostrictive, magnetostrictive, and shape memory materials are all capable of adding external mechanical energy to systems, they are useful when active control is appropriate. The ability of ER materials to add small amounts of mechanical energy to their surroundings through volumetric changes has been suggested in the literature [5,6]. This effect has not been widely studied, however, and in general ER materials are not considered when mechanical actuation of a system or structure is desired. Instead, ER materials are used in semi-active control applications where their controllable mechanical properties can be utilized to selectively tune overall system response.

Recalling that the rheological behavior of ER materials can be categorized into pre-yield and post-yield regimes, it is not surprising that applications investigated to date can be classified according to which of these regimes the ER material is exposed. ER material controllable devices for the most part are based on the controllable, post-yield, shear behavior of ER suspensions. Investigations of ER controllable devices such as valves, mounts, clutches, brakes, and dampers have by far dominated the ER materials applications research and development spectrum to date. Such devices are discussed in detail in the next section of this review. ER material adaptive structures are those that are made controllable through the incorporation of ER materials at selected locations. For common structural deformation ranges, it is likely that contained ER materials will remain in a pre-yield state. In this case, the tunable structural performance realized is based on controllable pre-yield ER material mechanical response. The progress of ER material adaptive structure research efforts is covered in a later section of this article.

ER MATERIAL CONTROLLABLE DEVICES

Post-Yield Rheological Behavior

ER material controllable devices have for the most part been designed to take advantage of the behavior of ER suspensions in shear loading situations. An idealized diagram of ER material shear response is shown in Figure 1, where shear stress is plotted as a function of both shear strain and shear strain rate. As shown in Figure 1(a) for small strain levels, ER materials remain in an unyielded state. Applications based on the pre-yield ER material behavior will be discussed later in this review. Controllable devices are based on the post-yield behavior depicted in Figure 1(b). For post-yield situations, the constitutive shear behavior of

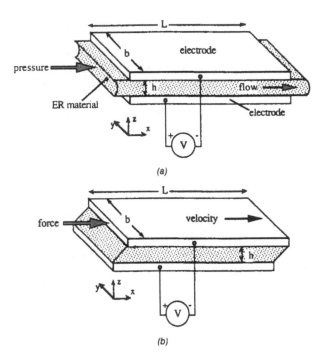

Figure 2. *Device component configurations: (a) fixed electrode and (b) sliding electrode.*

ER materials is often modeled using the Bingham plastic approximation

$$\tau = \tau_y + \eta \dot{\gamma} \qquad (1)$$

In this equation, τ_y is the dynamic yield stress, which for ER suspensions is a strong function of electric field. The plastic viscosity η of ER formulations is a weak function of electric field, and in the design of controllable devices, is often assumed constant.

Basic ER Material/Electrode Interaction Configurations

The primary device types that have been studied to date include valves, mounts, clutches, brakes, and dampers. The operation of each of these devices is based on one of the two fundamental ER material/electrode interaction configurations shown in Figure 2. These are called the fixed electrode configuration, shown in Figure 2(a), and the sliding electrode configuration, shown in Figure 2(b). Devices based on the fixed electrode configuration are designed so as to have stationary electrodes between which an ER material flows with a flow rate Q due to the existence of a pressure gradient Δp. The variable performance of such devices is realized by the control of the applied electric field E. The control of this parameter, along with the ER material shear properties and device geometry, governs the relationship between pressure drop and flow. Components based on the sliding electrode configuration are fabricated with parallel electrodes, at least one of which can move with a tangential speed S when subjected to a tangential shear force F. In such situations, the force/speed relationship is governed by the device geometry and the ER material shear properties.

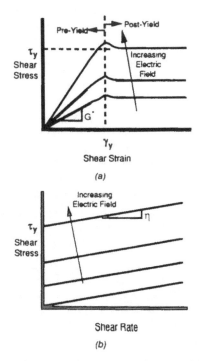

Figure 1. *Idealized constitutive shear behavior of ER materials.*

A basic understanding of the two fundamental ER material/electrode interaction situations can lead to an approximation of the performance of actual ER material controllable devices. Under the assumptions of idealized Bingham plastic ER material behavior and steady-state conditions, the relationship between flow rate and pressure drop for the fixed plate configuration, as well as that between applied force and electrode speed for the sliding electrode configuration, can be explicitly determined. Brief descriptions of the theoretical analyses that lead to these relationships follow.

Fixed Electrode Configuration

The basic understanding of the fixed electrode configuration depicted in Figure 2(a) dates back to a dissertation written by Phillips in 1969 [7,8]. In Figure 3, a section of a fixed electrode configuration is shown, with a small control volume element in the x-z plane expanded so as to illustrate the primary forces acting on the element. Assuming that the flow of the ER material is one-dimensional, and that body forces and convective effects are negligible, the application of momentum conservation to the situation depicted in Figure 3 leads to a first-order differential equation relating shear stress to the axial pressure gradient. The solution of this equation leads to the linear shear stress distribution across the electrode gap shown in Figure 4(a). This distribution is described by the equation

$$\tau(z) = -p' \left(z - \frac{h}{2} \right) \qquad (2)$$

with

$$p' = -\frac{dp}{dx} = \text{constant} \qquad (3)$$

The shear stress distribution is independent of the type of material that exists between the electrodes. When the mate-

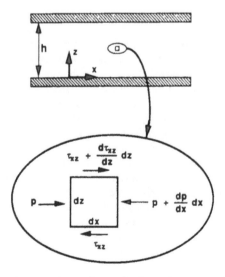

Figure 3. Control volume element for fixed electrode configuration analysis.

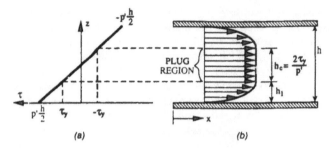

Figure 4. Stress and flow associated with fixed electrode configurations: (a) stress distribution and (b) velocity distribution.

rial has a characteristic yield stress, such as is the case with ER materials, no flow will result until the pressure gradient is increased to a level at which stresses larger than the yield stress exist. From Equation (2), it follows that the critical pressure gradient magnitude that must exist for the onset of flow is

$$\left| \left(\frac{dp}{dx} \right)_c \right| = \frac{2\tau_y}{h} \qquad (4)$$

When pressure gradient levels equal to or larger than this critical value exist, the material near the electrode walls where the largest shear stress values exist will yield and flow will result. Near the center region of the gap where the shear stress is identically zero, there will always be a section of unyielded material characterized by a uniform axial velocity similar to plug flow. This is shown in Figure 4(b), where a velocity profile across the gap of an ER material fixed electrode configuration is shown. The thickness of the core or plug region, h_c, depends on the yield stress and pressure gradient, as shown in the figure.

To completely understand the velocity profile that results in the gap, the detailed constitutive behavior of the flowing material must be known. If Bingham plastic behavior is assumed, a first-order differential equation for axial velocity results, which is:

$$\frac{du}{dz} = \frac{p'(h_1 - z)}{\eta} \qquad (5)$$

in which h_1 is the thickness of the sheared region on each side of the unyielded core. The solution of this equation subject to a no-slip boundary condition at the electrode surfaces leads to a complete description of the symmetric velocity distribution across the gap. Near an electrode wall, the velocity at any point in the sheared region is determined from the relation

$$u(z) = \frac{p'(2h_1 z - z^2)}{2\eta} = \left(\frac{p'h}{2\eta} - \frac{\tau_y}{\eta} \right) z - \frac{p'}{2\eta} z^2 \quad (0 \leq z \leq h_1) \qquad (6)$$

The unsheared core moves with a uniform velocity equal to

$$u_c = \frac{p'h_1^2}{2\eta} = \frac{1}{8p'\eta} (p'h - 2\tau_y)^2. \quad \left(h_1 \leq z \leq \frac{h}{2} \right) \qquad (7)$$

Equations (6) and (7), along with the symmetric nature of the flow, define a complete velocity profile such as that represented in Figure 4(b).

The volume flow rate, Q, can be directly calculated through the integration of the velocity profile across the gap. When this is done, the resultant flow rate per unit gap length is given by the relationship

$$\frac{Q}{b} = \frac{(p'h - 2\tau_y)^2(p'h + \tau_y)}{12p'^2\eta} \qquad (8)$$

Defining a dimensionless pressure drop P^* and yield stress T^* as follows,

$$P^* = \frac{bh^3p'}{12Q\eta}, \quad T^* = \frac{bh^2\tau_y}{12Q\eta} \qquad (9)$$

Equation (8) can be manipulated into an equivalent third-order form, which is

$$P^{*3} - (1 + 3T^*)P^{*2} + 4T^{*3} = 0 \qquad (10)$$

The solution of Equation (10) provides the desired relationship between pressure drop, flow rate, material properties, and geometry for fixed electrode ER material configurations. Because the equation is of cubic form, there are three roots for P^*. A classical algebraic analysis of the equation with a reasonable assumption of ER material property ranges leads to the unfortunate conclusion that all three of the P^* roots are real, and that the equation is irreducible. For actual ER material applications, however, it was concluded by Phillips that only one of these three roots makes physical sense [7]. This solution is illustrated in Figure 5, where the dimensionless pressure gradient is plotted as a function of the dimensionless yield stress.

The equation representing the interdependence of the dimensionless variables plotted in Figure 5 is quite complex. For this reason, more simplistic relations that approximate the exact behavior are often used to analyze ER material controllable devices. An equation that serves as an upper bound to the exact behavior is

$$P^* = 1 + 3T^* \qquad (11)$$

This relationship approximates the exact behavior of ER material fixed electrode situations to within ±5% for values of T^* less than 0.5. Conditions characterized by T^* values below this threshold are often referred to as "high flow" situations, since the ratio of yield stress to flow rate is relatively small. When written in terms of dimensional quantities, Equation (11) assumes the form

$$\Delta P = \Delta P_{HI} + \Delta P_{ER\,HI} = \frac{12\eta QL}{bh^3} + 3\frac{L}{h}\tau_y \qquad (12)$$

The first and second terms on the right hand side of this pressure drop relation are associated with viscous and yield effects, respectively.

At the other extreme of ER material fixed electrode configuration behavior is the "low flow" regime, characterized by values of T^* greater than 200. In such situations, the ratio of ER material yield stress to flow rate is large, and the dimensionless pressure gradient can be determined to within ±5% using the relation

$$P^* = \frac{2}{3} + 2T^* \qquad (13)$$

In dimensional form, the equation for pressure drop that is equivalent to Equation (13) is

$$\Delta P = \Delta P_{LI} + \Delta P_{ER\,LI} = \frac{8\eta QL}{bh^3} + 2\frac{L}{h}\tau_y \qquad (14)$$

where again the two contributions to pressure drop are associated with viscous and yield effects, respectively.

Graphical representations of Equations (11) and (13) are included in Figure 5, where it can be seen that the approximations are quite accurate over the ranges of T^* stated. In the analysis of actual ER material devices, the pressure drop contribution due to the existence of a yield stress often dominates. In many cases a constant, C, is introduced to allow the pressure drop associated with yield stress to be uniformly represented as

$$\Delta P_{LR} = C\frac{L}{h}\tau_y \qquad (15)$$

The numerical value for C ranges from 2 to 3 depending on the regime of the flow situation being studied.

Sliding Electrode Configuration

In sliding electrode situations, the force applied to an electrode is distributed evenly over the electrode surface, resulting in a uniform shear stress level at the ER material/electrode interface. The fundamental relationship between the force, F, interface shear stress, τ, and electrode surface area, A, is:

$$F = \tau A \qquad (16)$$

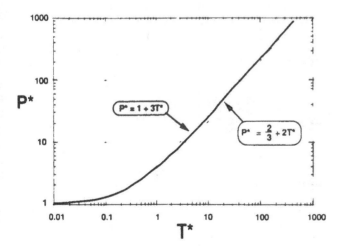

Figure 5. *Dependence of dimensionless pressure drop on dimensionless yield stress for fixed electrode ER material configurations.*

The shear stress τ is determined from the Bingham plastic constitutive relationship for the ER material, which was previously presented as Equation (1). If a linear velocity distribution across the ER material is assumed and S represents the velocity differential between the electrodes, the shear rate simply becomes

$$\dot{\gamma} = \frac{S}{h} \qquad (17)$$

Recognizing that the electrode area depicted in Figure 2(b) is $A = Lb$ and substituting this along with Equations (1) and (17) into Equation (16) leads to the desired force-speed relationship, which is

$$F = Lb\tau_y + \frac{SLb\eta}{h} \qquad (18)$$

The first term in this equation is the contribution to the transmitted force that is due to the yield stress of the ER material. The second term, which describes the force that would result if the material were Newtonian with a plastic viscosity equal to η, is called the viscous force term. Because of the uniform shear rate assumption presented as Equation (17), Equation (18) is completely valid only for planar electrode configurations such as that shown in Figure 2(b). In actual ER material applications research, however, Equation (18) has been used to approximate the performance of axisymmetric ER material controllable devices which have cylindrical rather than planar electrode surfaces.

ER Material Volume and Electrical Power Requirements

In addition to the mechanical performance of an ER material controllable device, two factors that are of significant importance are the amount of ER material needed and the electrical power required. In many cases, applications are specified in terms of a mechanical control ratio which relates the device performance with the electric field applied to that when the field is off. For devices based on the fixed electrode configuration, the mechanical control ratio is defined as the ratio of the pressure drop with the electric field applied to that obtained in the absence of an electric field. In terms of this control ratio, the volume of ER material required for a fixed electrode controllable device is

$$(\text{Volume})_f = \frac{12}{C^2} \frac{\eta}{\tau_v^2} Q \Delta P_{1R} \left(\frac{\Delta P_{1R}}{\Delta P_{1ll}} \right) \qquad (19)$$

If the electrical current per unit area of electrode surface is identified as J, the electrical power required for a fixed-electrode controllable ER material device is:

$$(\text{Power})_f = \frac{12}{C^2} \frac{\eta}{\tau_y^2} Q \Delta P_{1R} \left(\frac{\Delta P_{1R}}{\Delta P_{1ll}} \right) EJ \qquad (20)$$

It is a general rule that in designing an ER material device one attempts to minimize both the ER material volume and the electrical power required. From Equations (19) and (20), it can be seen that from a materials perspective, the best ER formulations for fixed-electrode controllable de-

vices are those characterized by a high yield stress and a low viscosity.

A mechanical control ratio that is appropriate for devices based on the sliding electrode configuration is the ratio of transmitted force with the electric field applied to that with the field off. In terms of this ratio, the volume of ER material required for such a device is

$$(\text{Volume})_s = \frac{\eta}{\tau_v^2} SF_{1R} \left(\frac{F_{1R}}{F_o} \right) \qquad (21)$$

The electrical power required to operate a sliding electrode ER material device is

$$(\text{Power})_s = \frac{\eta}{\tau_v^2} SF_{1R} \left(\frac{F_{1R}}{F_o} \right) EJ \qquad (22)$$

From Equations (21) and (22), it can be concluded that like fixed-electrode devices, optimal sliding electrode device performance is realized when an ER material is selected so as to have a low viscosity and a high dynamic yield stress [9–11].

Range of Applicability of the Basic Theories

The theories developed in the previous sections are completely valid only when parallel planar electrode configurations exist and all of the material and kinematic assumptions made are valid. When such conditions exist, the relationships resulting from the analyses can be applied with great confidence. In situations where one or more of the fundamental assumptions is not valid, these relationships can be used only to approximate device behavior. In many cases, the resultant approximations turn out to be quite accurate.

When the analysis of a particular ER material controllable device does require a deviation from the basic assumptions, the modification can be classified as either geometrical, material, or dynamic. Geometrical modifications to the theory are in some cases necessary when electrodes are not planar parallel surfaces, as assumed above. A very common electrode configuration utilized in ER material devices, for example, is a concentric cylinder configuration. Material modifications are required when the constitutive post-yield shear behavior of an ER material to be utilized deviates significantly from idealized Bingham plastic form over the range of operational shear rates relevant to a target device. In certain cases, accounting for the slightly non-Bingham nature of certain ER formulations can lead to a much better understanding of actual controllable device performance. Lastly, dynamic theoretical modifications are required when transient behavior and/or electrode/ER material movement patterns other than those discussed are important. Also, when high frequency switched power supplies and/or AC based ER materials are employed, it is important to realize that electrical capacitance effects must also be considered.

Theoretical improvements such as those alluded to above that go beyond the basic theories in order to better understand particular classes of controllable devices do exist. The modifications that are needed or have been implemented in certain instances range in complexity, and are considered beyond the scope of this introductory review. The basic theories presented serve the purpose of describing how these controllable materials are generally used to produce controllable devices. With this understanding, the following review of developments in each of the device areas will be better appreciated.

Controllable Valves and Shakers

Controllable valves are among the original classes of ER material devices identified and investigated by Winslow [1]. ER valves are simply devices containing fixed electrode configurations similar to that depicted in Figure 2(a). Provided that the material flowing through the device is an ER suspension, the pressure drop across and flow rate through the device can be controlled by the application of an electric field. Benefits of ER valves include fast response time and an absence of mechanical moving parts. The theoretical understanding of ER valves as developed by Phillips was previously summarized [7]. A model for the onset of flow through ER valves employing a fundamental electrical approach was developed by Arguelles et al. [12]. Several investigators have constructed and tested ER material controllable valves experimentally [13–19]. Valve pressure drop levels as high as 6.9 MPa (1000 psi) have been achieved, and modulation frequencies on the order of several hundred hertz have been realized [13].

ER material valves have been proven very useful. In fact, several types of controllable devices have been developed as a result of the incorporation of ER valves into larger systems. Controllable shakers or vibrators, for example, have been developed through the assembly of ER valves into the hydraulic equivalent of an electrical Wheatstone bridge arrangement. An oscillatory vibrational output is realized by the modulation effect of such arrangements on an external hydraulic pressure source. An analytical study of ER shakers, which concluded that high force output and wide operational bandwidth were possible, was completed over two decades ago by Phillips and Auslander [8]. Prototype shakers have been evaluated on several occasions. One of the earlier experimental efforts was performed by Eige and Sevy, who were able to vibrate an 11 N (2.5 lb.) load with a maximum peak to peak acceleration of 29 g at frequencies up to 200 Hz. Their design was based on the placement of four distinct ER valves in hydraulic lines feeding a traditional piston-cylinder arrangement [20]. A somewhat different ER shaker was investigated by Strandrud several years later [15,16]. He constructed a piston-cylinder arrangement with two ER valves attached to the cylinder and two incorporated as piston heads on an output shaft. With his arrangement, he observed response times of less than one millisecond and was able to generate sinusoidal force patterns with amplitudes as high as 9880 N (2200 lb.) at frequencies as high as 1000 Hz. Some years later, Brooks contributed to the advancement of ER material shakers by developing and analyzing a prototype vibrator with a design payload of 1000 N and a target operational frequency range of 1–1000 Hz [13]. Stangroom [17] has also worked with ER material shakers, and several groups are known to be continuing research efforts in this area at the present time [21,22].

Controllable Machinery and Engine Mounts

Another class of devices that have been made controllable through the incorporation of ER material valve technology is machinery and engine mounts. A schematic diagram of a possible controllable mount configuration is shown in Figure 6. Traditional fluid filled mounts, which are configured much like that shown in Figure 6, are designed to have fluid inertia track characteristics, as well as top and bottom compliance values appropriate for specific force transmission applications. The number, size, and shape of fluid inertia tubes are application specific, and in traditional mounts are fixed once the mount is fabricated. The introduction of ER valves as inertia track components brings with it a capability for temporal fluid inertia variation, and as a result overall mount performance control.

One of the first investigations of ER material based controllable mounts was performed by Duclos in 1987 [14]. He completed a coupled analytical and experimental study encompassing individual valve design, prototype mount design, mount performance simulation, and experimental testing. A multiple fluid inertia track design was selected so as to maintain the high static stiffness of traditional mounts and produce a low dynamic stiffness over a target frequency range of 20–40 Hz. The prototype mount was tested at specific frequencies within the target range and also under conditions where input frequency sensing and automatic computer control was utilized. In all instances, the desired wide bandwidth, low dynamic stiffness performance was achieved. Results of subsequent efforts related to this study have been reported on several occasions [9,10,23].

Controllable ER material engine mounts have also been investigated by Ushijima et al. [24]. These authors experimentally studied a base excited vibration system where the force is transmitted through an ER controllable mount over the frequency range of 10–50 Hz. A significant range of controllability was observed, and simulations which illustrated the feasibility of utilizing ER mounts in semi-active control systems were developed.

A third group that has recently studied ER controllable mounts is Petek and coworkers [25]. This group focused on a 1–50 Hz frequency range and developed a prototype mount to have two states; namely an *off* state characterized by very low damping and a low dynamic spring rate, and an *on* state characterized by very high damping and a larger dynamic spring rate. The mount was tested in an on-off

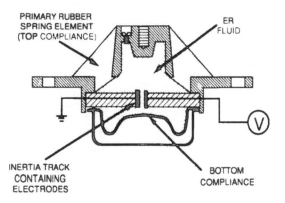

Figure 6. *Possible ER machinery or engine mount configuration.*

feedback control environment with a test apparatus custom designed to simulate anticipated conditions. Significant control over mount transmissibility both near and away from resonance conditions was achieved. An additional feature of this investigation was the testing of the prototype mount at 0°C and 100°C, where positive control capabilities were still observed.

A final investigation that should be mentioned in the area of ER controllable mounts was performed by Oppermann et al. [26]. As part of this investigation, an ER material mount was constructed and tested over a temperature range of −30°C to 110°C. Endurance tests that were performed indicated that the mount could withstand over 20 million load changes at a temperature of 110°C for a period of several hundred hours without being damaged. A novel component of this investigation was the use of the electrical response of the ER material to sense velocities within the mount.

Controllable Clutches and Brakes

Like ER valves, controllable ER clutches and brakes were originally investigated by Winslow and reported in his 1949 disclosure [1]. Little was accomplished in further developing these devices over the next three decades. However, since the early 1980s, ER clutches and brakes have received much attention. ER clutches are based on the sliding plate basic device component configuration discussed earlier. They can be configured in either concentric cylinder or parallel disk configurations, as shown in Figure 7.

In 1983, Sproston et al. performed an experimental investigation of the torque transmission across a single element parallel disk clutch [27]. A prototype clutch mechanism was designed and tested at a constant rotation rate of 60 rpm. ER material gap thickness and applied voltage were varied, and a linear relationship between torque transmission and applied voltage was observed. Several years later, a follow-up study was performed by this same research group [28]. An improved single element parallel disc clutch assembly was constructed, and torque transmission at rotational speeds as high as 1200 rpm were experimentally studied at various applied voltage levels. Special attention was given to the temperature of the ER materials in the device, and the potential negative effects of electrical and viscous material heating were studied. For the materials and conditions studied, peak torque transmission values were observed when the ER material temperature was in the range of 50–55°C.

A study of ER controllable brakes was also performed by Seed et al. [29]. These investigators selected design target specifications that included the control of torque within a 0.2–2.2 N·cm range over a rotation rate range of 0–4300 rpm. They designed and constructed a concentric cylinder type configuration that closely approached their design specifications during testing. Like the group led by Stevens and Sproston, Seed et al. observed variations in brake mechanical performance due to viscous heating of the ER material in the device.

More recently, Stangroom has studied ER controllable clutches. This was done in the context of developing a specific product for the control of wire tension during manufacturing processes [17,30]. He selected manufacturing as a target area for ER devices because of the relatively narrow operation temperature range characteristic of factory environments. To satisfy the maximum torque requirement of greater than 6 N·m, Stangroom developed a novel design that utilized a parallel disk ER clutch to modulate the tension on a friction band that actually carried the primary load.

ER clutches and brakes have also been studied extensively by Carlson and coworkers [11,31]. These investigators focused primarily on the development of prototype clutches of the parallel disk form. A key contribution from this effort was the development and partial validation of design equations with which controllable ER clutch applications can be evaluated. The designs developed by Carlson et al. were unique in that magnetic coupling arrangements were utilized, thus allowing for the elimination of traditionally problematic dynamic seals. A prototype ER clutch incorporating magnetic coupling was tested at torque levels up to 0.07 N·m.

In a related effort, Carlson collaborated with Colvin and investigated the usage of ER controllable clutches in fall intervention systems for elderly individuals [32]. It was desired to provide up to a 225 N (50 lb.) braking force to a 450 N (100 lb.) elderly patient that was falling from an original standing position. A parallel disk prototype brake incorporating magnetic coupling and capable of providing up to 5.0 N·m of torque was fabricated and tested using an overhead tether system and feedback control. Tests performed with dead weights verified the capabilities of the system to provide the desired braking force, but tests performed with an actual manikin produced torque transients that exceeded the capacity of the ER brake. It was concluded that a larger design safety margin would be required to successfully meet the proposed application requirements.

Finally, a current ER material controllable clutch effort that should be recognized is that of Johnson et al. [33]. His investigation includes the evaluation of ER clutches at high speeds. Although the shear rates to which ER materials are exposed tend to high in this application, positive control capabilities have been observed.

Controllable Dampers

Both of the basic configurations discussed earlier in this review can be utilized to create controllable ER material mechanical dampers. Example configurations for the two resultant device types, known as fixed plate and sliding plate dampers, respectively, are shown in Figure 8. In a fixed plate damper, the damping force on a piston is realized by

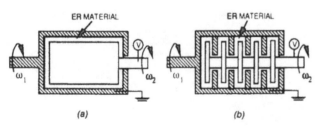

Figure 7. ER clutch configurations: (a) concentric cylinder and (b) parallel disk.

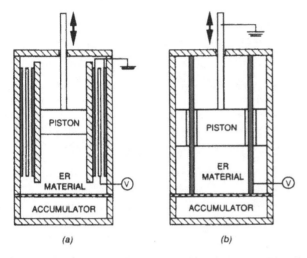

Figure 8. *ER damper configurations: (a) fixed plate and (b) sliding plate.*

the control of the pressure drop across valve-like channels through which the ER material is forced to flow. In a sliding plate damper, the damping force originates from the controlled shear resistance between the moving piston, which acts as one electrode, and adjacent parallel surfaces, which remain motionless and act as the other electrode. In both types of damping devices, accumulators are often required to account for volume variations due to the travel of the piston rod into and out of the damper cavity.

An experimental evaluation of ER dampers was performed in the mid 1980s by Stevens et al. [34]. During the first phase of this investigation, a sliding plate damper was constructed and tested in a linear second-order spring-mass-damper system. The system was excited harmonically over a frequency range of 0.5 to 5.0 Hz and the steady-state amplitude of motion for the mass was measured at various levels of applied voltage. A significant and controllable variation in mass amplitude attenuation was observed. As anticipated, the largest attenuation control range resulted at the system resonance frequency of approximately 2.5 Hz.

In a subsequent analytical investigation, this same group of investigators developed a model of the previously discussed ER damper system [35]. The model, which included the effects of both Coulomb and viscous friction, allowed for direct comparisons between analytical predictions and experimental observations. Although the model was determined to overestimate system damping near resonance, reasonable qualitative agreement between theory and experimentation resulted.

Another experimental and analytical study of ER material dampers was reported by Shulman et al. in 1987 [36]. An analytical model that incorporated the visco-elastic nature of ER materials was developed for isolation systems in which a mass was connected to a disturbance platform by a parallel spring ER damper arrangement. The model was utilized to study both free and forced vibrations. In a corresponding experimental effort, an actual isolation system that incorporated four sliding plate ER dampers was developed. Small displacement amplitude tests were performed over a 4-40 Hz frequency range for both free and forced

oscillation conditions. Through the observed isolation characteristics, the importance of both viscous and elastic material behavior was demonstrated. As much as a six-fold variation in system damping properties was observed.

An interesting analytical study focusing on the feasibility of ER dampers as aircraft landing gear components was recently performed by Bhadra et al. [37]. The landing sequence of typical aircraft was analyzed and an optimal damping strategy proposed. The requirements of ER dampers for optimized landings on smooth and rough terrain runways were discussed. It was concluded that current ER materials do not meet the necessary requirements, but that future improved formulations perhaps will.

In addition to his work with other ER devices, Brooks has investigated prototype ER dampers for several specific applications [13]. During one phase of his study, he constructed a rather large damping strut designed to reduce helicopter vibrations. The damper that was constructed contained over 3700 cm² of sliding electrode interaction area. When operated at a field level of 3.7 kV/mm, this damper required over 50 W of electrical energy. The application specifications included a required dynamic load of 2000 N over a frequency range of 0-20 Hz. The strut was tested in a laboratory over a frequency range of 0-150 Hz, with the required specifications being met. Testing on an actual helicopter was never performed. Brooks also developed a sliding plate torsion bar damper for off-road heavy vehicle applications. This damper was tested over a frequency range of 0-2.0 Hz and found to controllably support approximately 1000 N·m of torque.

ER controllable dampers have also been developed at Lord Corporation by Duclos and coworkers [9,23,31]. Design equations based on a sliding electrode configuration were developed and a corresponding prototype ER damper was constructed. Theoretical and experimental damping forces were compared for steady unidirectional motion situations. With seal and spacer frictional considerations taken into account, a reasonably good level of correlation between predictions and observations was found.

Additional Controllable Devices and Application Requirements

In addition to those discussed above, many other interesting controllable devices based on the unique behavior of ER suspensions have been proposed. The device classes discussed above are simply those that by far have received the most attention. Proposed controllable devices that are not discussed in this review range from peristaltic pumps to penile implants [6,38].

Important additional contributions to ER controllable device technology have been made by investigators who have sought to define device requirements necessary for commercially attractive applications. In 1988, Duclos evaluated ER devices for automotive applications [10]. He concluded from a purely mechanical standpoint that ER engine mounts were feasible at the present time and with modest improvement in ER material properties, several damper applications would become practical. More recently, Hartsock et al. completed a similar evaluation of potential ER material based automotive devices [39]. These authors evaluated damper, mount, and clutch capabilities with current, proposed near-term, and possible long-term ER materials. They concluded that only engine mounts were feasible with currently realized ER material properties. With improved near-term ER formulations, shock absorbers became feasible, and with significant long-term improvements, all evaluated devices were feasible.

ER MATERIAL ADAPTIVE STRUCTURES

Investigations concerning adaptive structures technology have only recently been initiated. A concise review of several pertinent early advancements in this area has recently been prepared by Wada et al. [40]. The concept of using ER materials in adaptive structures is quite new, and can be traced back to a patent issued to Carlson et al. [41]. Unlike ER controllable devices, adaptive structures are primarily based on the controllability of the pre-yield rheology of ER suspensions. By incorporating a material with known controllable rheology into an otherwise passive structure, the response of the entire composite system becomes tunable. It thereby becomes possible for the structure to adapt to a variable environment in the interest of continuously optimizing performance.

There are no theoretical rules for how ER materials are to be incorporated into structures, but to date, two classes of basic configurations have been studied. These can be referred to as a shear configuration and an extensional configuration. ER adaptive structures based on these configurations are illustrated in Figure 9. Since axial flexure of the beam-like structure shown in Figure 9(a) causes the ER material in the structure to primarily undergo shear deformation, it is entitled "shear configuration". Likewise, flexure of the structure shown in Figure 9(b) will expose the ER material to primarily extensional and compressive stresses aligned with suspended particle chains; thus, this configuration is given the name "extensional configuration". A summary of investigations focusing on ER adaptive structures based on these configurations follows.

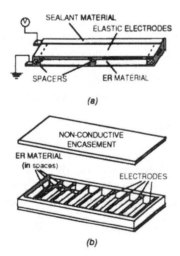

Figure 9. Basic ER adaptive structure components: (a) shear configuration and (b) extensional configuration.

Shear Configurations

A theoretical and experimental study of ER adaptive structures based on shear configurations was performed by Coulter et al. [9,23,31,42]. Sandwich beams similar to that shown in Figure 9(a) were fabricated and placed in a state of continuous flexural vibration over a frequency range of 0–200 Hz. Under such conditions, several modes of vibration were excited simultaneously, and the response of each beam was monitored. Varying response was observed as the level of electric field applied to the ER material within the structures was changed. Structural natural frequencies for each of the vibrational modes studied were observed to increase linearly with applied electric field. Although loss factors associated with each mode were also seen to increase with electric field level, the form of this increase varied depending on the structure tested. A model based on analyses commonly applied to passive viscoelastic material shear damping treatments was proposed, and direct comparisons between theoretical predictions and experimental observations were made. Qualitative agreement between theory and observation resulted, but the quantitative correlation was not definitive. It was suggested that modifications to the model to accommodate non-ER damping layer components and more reliable ER complex modulus data are needed before a complete quantitative evaluation of the modelling concept can be performed.

Another group to study shear configuration based ER adaptive structures was that of Gandhi et al. [43–47]. This group experimentally studied a variety of shear configuration based structures in vibrational environments. They observed significant changes in both natural frequency and damping ratio with applied electric field in all cases. In an attempt to simulate the observed controllable behavior, a three-dimensional finite-element model was developed. Simulations were performed using empirically determined structural damping and stiffness constants, and theoretical vibrational attenuation levels were predicted.

A third effort related to structures containing shear configurations was performed by Choi et al. [48]. During this experimental study, shear configuration based beam specimens were fabricated and tested in cantilever type free vibration situations. Values for the complex structural rigidity of the entire composite system were determined from the observed response. Significant controllability of system vibrational response was observed.

Extensional Configurations

A theoretical and experimental investigation of extensional configuration based ER adaptive structures was performed by Margolis and Vahdati [49,50]. By focusing on a single unit cell geometry, these researchers developed a modified Timoshenko beam model which incorporates controllable ER material response. With this model, simulations of the flexural vibration of multi-cell beam structures were performed, and control approaches for variation of the applied electric field were developed. Simulations were used to determine the control approaches that would provide optimum damping of these adaptive structures. The model was partially validated for static deflection situations. Experimentation was performed to study the static and dynamic response of a single unit test cell consisting of an ER material held between two parallel metal electrodes. One of the electrodes was oscillated about an inplane central axis while the other was held motionless. Significant stiffening of the test cell upon the application of an electric field was observed, and for the static deformation case, the controllable spring capabilities that can be realized with ER materials was demonstrated.

In a study related to that of Margolis and Vahdati, an extensional configuration based ER adaptive structure was constructed and tested by Coulter [51]. The structure contained fifty axial ER unit cells that were enclosed and held into position by a molded viscoelastic polymer shell. The structure was tested in a cantilevered arrangement over a frequency range of 0–200 Hz. Six vibrational modes were observed within this bandwidth. Significant natural frequency and modal damping increases were observed as applied electric field was increased.

A final experimental study performed with an ER material adaptive flexible cylinder containing nearly 200 ER material/electrode extensional type unit cells was performed by Austin [52]. In this study, the structure was excited into longitudinal vibration using piezoceramic actuators. Natural frequency changes and as much as a three-fold increase in logarithmic decrement was observed.

SUMMARY AND CONCLUSIONS

As is evident from the contents of this article, a significant amount of international effort has been directed towards the development of ER material applications. The controllable rheological nature of these materials has been evaluated for a wide range of application concepts. Significant advancements in the development and understanding of ER material controllable devices have been realized, with the primary proposed device types being valves, mounts, clutches, and dampers. In addition, during the past several years the new potential ER application area of adaptive structures has emerged. Proof of concept studies in this area has been completed, but the theoretical understanding of ER adaptive structures requires further development.

As a result of many of the studies reviewed in this article, additional critical parameters related to ER applications have been identified. Further investigations are required to bring about a clear understanding of some of the issues that have surfaced. The suggested near-term market for ER material based systems is believed to be quite large, and a concentrated and cohesive effort by the international electrorheological materials community is going to be required before this market can begin to be tapped.

ACKNOWLEDGEMENT

Funding for J. P. Coulter's participation in the present study was provided by The National Science Foundation in the form of Grant No. MSS-9110909, and the U.S. Army Research Office in the form of Grant No. DAAL03-92-G-0388.

NOMENCLATURE

A	electrode surface area
b, h, L	configuration dimensions
C	flow regime constant
E	applied electric field
F	force
h_c	core region thickness
h_1	sheared region thickness
J	electrical current density
p	pressure
p'	pressure drop ($= -dp/dx$)
$P*$	dimensionless pressure drop
Q	flow rate
S	electrode speed
$T*$	dimensionless yield stress
u	velocity
V	applied voltage
x, y, z	Cartesian directional coordinates
γ	shear strain
η	plastic viscosity
τ	shear stress
τ_v	dynamic yield stress

REFERENCES

1. Winslow, W. M. 1949. "Induced Fibration of Suspensions", *Journal of Applied Physics*, 20(12):1137–1140.

2. Weiss, K. D., J. P. Coulter and J. D. Carlson. 1993. "Material Aspects of Electrorheological Systems", *Journal of Intelligent Material Systems and Structures*, 4(1):13–34.

3. "Smart Fluids—New Route to Advanced Hydraulic Systems/Devices", Emerging Technologies Report No. 35.

4. Goldstein, G. 1990. "Electrorheological Fluids: Applications Begin to Gel", *Mechanical Engineering*, 112:48–52.

5. Gorodkin, R G., Y. V. Korobko, G. M. Blokh, V. K. Gleb, G. I. Sidorova and M. M. Ragotner. 1979. "Applications of the Electrorheological Effect in Engineering Practice", *Fluid Mechanics Soviet Research*, 8(4):48–61.

6. Shulman, Z. P., R. G. Gorodkin, E. V. Korobko and V. K. Gleb. 1981. "The Electrorheological Effect and Its Possible Uses", *Journal of Non-Newtonian Fluid Mechanics*, 8:29–41.

7. Phillips, R W. 1969. "Engineering Applications of Fluids with a Variable Yield Stress", Ph.D. dissertation, The University of California at Berkeley.

8. Phillips, R W. and D. M. Auslander. 1971. "The Electroplastic Flow Modulator", *Proceedings of Flow: Its Measurement and Control in Science and Industry, May 9–14, 1971, Pittsburgh*, pp. 1243–1251.

9. Coulter, J. P. and T. G. Duclos. 1989. "Applications of Electrorheological Materials in Vibration Control", in *Proceedings of the Second International Conference on Electrorheological Fluids*, J. D. Carlson, A. F. Sprecher and H Conrad, eds., Lancaster, PA: Technomic Publishing Company, Inc., pp. 300–325.

10. Duclos, T. G. 1988. "Design of Devices Using Electrorheological Fluids", Paper no. 881134, The Society of Automotive Engineers, Inc., Detroit, Michigan.

11. Carlson, J. D. and T. G. Duclos. 1989. "ER Fluid Clutches and Brakes—Fluid Property and Mechanical Design Considerations", in *Proceedings of the Second International Conference on Electrorheological Fluids*, J. D. Carlson, A. F. Sprecher and H. Conrad, eds., Lancaster, PA: Technomic Publishing Company, Inc., pp. 353–367.

12. Arguelles, J., H. R. Martin and R. Pick. 1974. "A Theoretical Model for Steady Electroviscous Flow between Parallel Plates", *Journal of Mechanical Engineering Science*, 16(4):232–238.

13. Brooks, D. A. 1989. "Devices Using Electro-Rheological Fluids", in *Proceedings of the Second International Conference on Electrorheological Fluids*, J. D. Carlson, A. F. Sprecher and H. Conrad, eds., Lancaster, PA: Technomic Publishing Company, Inc., pp. 371–401.

14. Duclos, T. G. 1987. "An Externally Tunable Hydraulic Mount Which Uses Electro-Rheological Fluid", Paper no. 870963, The Society of Automotive Engineers, Inc., Detroit, Michigan.

15. Strandrud, H. T. 1966. "A Broadband Hydraulic Vibration Exciter", *The Shock and Vibration Bulletin*, 35(2):157 163.

16. Strandrud, H. T. 1966. "Electric-Field Valves Inside Cylinder Control Vibrator," *Hydraulics and Pneumatics*, pp. 139–143.

17. Scott, D. 1984. "Amazing Hardening Fluid Opens a New World of Hydraulic Devices," *Popular Science*, pp. 82–85.

18. Peel, D. J. and W. A. Bullough. 1989. "Miscellaneous Electro-Rheological Phenomena, Part III", in *Proceedings of the Second International Conference on Electrorheological Fluids*, J. D. Carlson, A. F. Sprecher and H. Conrad, eds., Lancaster, PA: Technomic Publishing Company, Inc., pp. 141–158.

19. Whittle, M., R. Firoozian and W. A. Bullough. 1989. "Analysis of Electro-Rheological Valve Responses: Steady, Frequency, and Time Domain", in *Electrorheological Fluids: Mechanisms, Properties, Applications*, R. Tao. ed., London: World Scientific Publishing Co., Inc., pp. 343–366.

20. Eige, J. J. and R. W. Sevy. 1962. "A Vibration-Shock Exciter Using Direct Electric-Field Modulation of Hydraulic Power", *Shock, Vibration and Associated Environments, Part V*, Washington, D.C.: Office of the Secretary of Defense, Research and Engineering Division, pp. 12–16.

21. Brooks, D. 1992. "Design and Development of Flow Based ER Devices", in *Electrorheological Fluids: Mechanisms, Properties, Applications*, R. Tao. ed., London: World Scientific Publishing Co., Inc., pp. 367–397.

22. Lou, Z., R. Ervin and F. E. Filisko. 1992. "Behavior of an ER Valve Configured as a Four-Arm Bridge", in *Electrorheological Fluids: Mechanisms, Properties, Applications*, R. Tao. ed., London: World Scientific Publishing Co., Inc., pp. 398 423.

23. Duclos, T. G., J. P. Coulter and L. R. Miller. 1989. "Applications for Smart Materials in the Field of Vibration Control", in *Smart Materials, Structures, and Mathematical Issues*, C. A. Rogers, ed., Lancaster, PA: Technomic Publishing Co., Inc., pp. 132 146.

24. Ushijima, T., K. Takano and H. Kojima. 1988. "High Performance Hydraulic Mount for Improving Vehicle Noise and Vibration", Paper no. 880073, The Society of Automotive Engineers, Inc., Detroit, Michigan.

25. Petek, N. K., R. J. Goudie and F. P. Boyle. 1989. "Actively Controlled Damping in Electrorheological Fluid-Filled Engine Mounts", in *Proceedings of the Second International Conference on Electrorheological Fluids*, J. D. Carlson, A. F. Sprecher and H. Conrad, eds., Lancaster, PA: Technomic Publishing Company, Inc., pp. 409 418.

26. Oppermann, G., G. Penners, M. Schulze, G. Marquardt, R. Flindt and T. H. Naumann. 1989. "Applications of Electroviscous Fluids As Movement Sensor Control Devices in Active Vibration Dampers", in *Proceedings of the Second International Conference on Electrorheological Fluids*, J. D. Carlson, A. F. Sprecher and H. Conrad, eds., Lancaster, PA: Technomic Publishing Company, Inc., pp. 287 299.

27. Sproston, J. L., N. G. Stevens and I. M. Page. 1983. "An Investigation of Torque Transmission Using Electrically Stressed Dielectric Fluids", in *Proceedings of Electrostatics 1983*. Oxford: The Institute of Physics, pp. 53 58.

28. Stevens, N. G., J. L. Sproston and R. Stanway. 1988. "An Experimental Study of Electro-Rheological Torque Transmission", *ASME Journal of Mechanisms, Transmissions, and Automation in Design*, 110:182 188.

29. Seed, M., G. S. Hobson, R. C. Tozer and A. J. Simmonds. 1986. "Voltage-Controlled Electrorheological Brake", *Proceedings of the IASTED International Symposium on Measurements, Processes, and Controls*, Sicily, September 3, pp. 280–284.

30. Stangroom, J. E. 1989. "Tension Control Using ER Fluids A Case Study", in *Proceedings of the Second International Conference on*

31. *Electrorheological Fluids*, J. D. Carlson, A. F. Sprecher and H. Conrad, eds., Lancaster, PA: Technomic Publishing Company, Inc., pp. 419–425.

31. Duclos, T. G., J. D. Carlson, M. J. Chrzan and J. P. Coulter. 1992. "Electrorheological Fluids—Materials and Applications", in *Intelligent Structural Systems*, H. S. Tzou and G. L. Anderson eds., Boston Kluwer Academic Publishers, pp. 213–242

32. Colvin, D. P. and J. D. Carlson. 1989. "Control of a Fail-Safe Tether Using an ER-Fluid Brake", in *Proceedings of the Second International Conference on Electrorheological Fluids*, J. D. Carlson, A. F. Sprecher and H. Conrad, eds., Lancaster, PA: Technomic Publishing Company, Inc., pp. 426–436.

33. Johnson, A. R., W. A. Bullough, R. Firoozian, A. Hosseini-Sianaki, J. Makin and S. Xiao. In press. "Testing on a High Speed Electro-Rheological Clutch", in *Electrorheological Fluids: Mechanisms, Properties, Applications*, R. Tao. ed., London: World Scientific Publishing Company, pp. 424 441.

34. Stevens, N. G., J. L. Sproston and R. Stanway. 1984. "Experimental Evaluation of a Simple Electroviscous Damper", *Journal of Electrostatics*, 15:275–283.

35. Stanway, R., J. L. Sproston and N. G. Stevens. 1987. "Non-Linear Modelling of an Electro-Rheological Vibration Damper", *Journal of Electrostatics*, 20:167–184

36. Shulman, Z. P., B. M. Khusid, E. V. Korobko and E. P. Khizhinsky. 1987. "Damping of Mechanical-Systems Oscillations by a Non-Newtonian Fluid with Electric-Field Dependent Parameters", *Journal of Non-Newtonian Fluid Mechanics*, 25:329 346.

37. Bhadra, D. K., C. R. Harder and W. B. Thompson. 1989. "Electroviscous Damping for Landing Aircraft", in *Proceedings of the Second International Conference on Electrorheological Fluids*, J. D. Carlson, A. F. Sprecher and H. Conrad, eds., Lancaster, PA: Technomic Publishing Company, Inc., pp. 402–408.

38. Rudloff, D. A. C. 1987. "Penile Implant", U.S. Patent 4,664,100, Issued May 12, 1987.

39. Hartsock, D. L., R. F. Novak and G. J. Chaundy. 1991 "ER Fluid Requirements for Automotive Devices", *The Journal of Rheology*, 35(7):1305–1326.

40. Wada, B. K., J. L. Fanson and E. F. Crawley. 1990. "Adaptive Structures", *Mechanical Engineering*, 12:41 46

41. Carlson, J. D., J. P. Coulter and T. G. Duclos 1990. "Electrorheological Fluid Composite Structures", U.S. Patent #4,923,057, Class Subclass 188 378, filed September 20, 1988, granted May 8, 1990

42. Coulter, J. P., T. G. Duclos and D. N. Acker. 1989. "The Usage of Electrorheological Materials in Viscoelastic Layer Damping Applications", in *Proceedings of Damping 89*. Wright Patterson Air Force Base, Ohio: Wright Aeronautical Laboratories CAAI-CAAI7.

43. Gandhi, M. V., B. S. Thompson and S. B. Choi 1989 "A New Generation of Innovative Ultra-Advanced Intelligent Composite Materials Featuring Electro-Rheological Fluids An Experimental Investigation", *Journal of Composite Materials*, 23 1232 1255

44. Gandhi, M. V. and B. S. Thompson 1989 "A New Generation of Revolutionary Ultra-Advanced Intelligent Composite Materials Featuring Electro-Rheological Fluids", in *Smart Materials, Structures, and Mathematical Issues*, C. A. Rogers, ed., Lancaster, PA Technomic Publishing Co. Inc., pp. 63–68.

45. Gandhi, M. V and B. S. Thompson 1989 "Dynamically-Tunable Smart Composites Featuring Electro-Rheological Fluids", in *Proceedings of the Fiber Optic Smart Structures and Skins II* E. Udd. ed., Boston: The Society of Optical Engineers, pp. 294 304

46. Gandhi, M. V., B. S. Thompson and S. B. Choi 1989 "A Proof-of-Concept Experimental Investigation of a Slider-Crank Mechanism Featuring a Smart Dynamically Tunable Connecting Rod Incorporating Embedded Electro-Rheological Fluid Domain", *Journal of Sound and Vibration*, 135(3) 511–515

47. Choi, S. B., B. S. Thompson and M. V. Gandhi 1989 "An Experimental Investigation on the Active-Damping Characteristics of a Class of Ultra-Advanced Intelligent Composite Materials Featuring Electro-Rheological Fluids", in *Proceedings of Damping 89*, Wright Patterson Air Force Base, Ohio Wright Aeronautical Laboratories CAC1-CAC14

48. Choi, Y., A. F. Sprecher and H. Conrad. 1990. "Vibration Characteristics of a Composite Beam Containing an Electrorheological Fluid", *Journal of Intelligent Material Systems and Structures*, 1(1): 91–104.

49. Margolis, D. L. and N. Vahdati. 1989. "The Control of Damping in Distributed Systems Using ER-Fluids", in *Proceedings of the Second International Conference on Electrorheological Fluids*, J. D. Carlson, A. F. Sprecher and H. Conrad, eds., Lancaster, PA: Technomic Publishing Company, Inc., pp. 326–348.

50. Vahdati, N. 1989. "Electrorheological Fluid Modeling for Controlled Damping of Distributed Systems", Ph.D. dissertation, The University of California at Davis.

51. Coulter, J. P. 1991. "Applications of Electrorheological (ER) Materials in Intelligent Devices and Structures—Lecture Notes", in *Intelligent Material Systems and Structures*, C. A. Rogers, ed., Lancaster, PA: Technomic Publishing Co., Inc.

52. Austin, S. A. 1992. "The Vibration Damping Effect of an Electrorheological Fluid", *Journal of Rheology*, submitted for publication.

Applicability of Simplified Expressions for Design with Electro-Rheological Fluids

Douglas Brooks

Advanced Fluid Systems Limited
10/14 Pensbury Industrial Estate
Pensbury Street
London SW8 4TJ
England

ABSTRACT: There are three key factors that enable electro-rheological (ER) fluids to be related to engineering hardware: *off-state*, which is characterised by the suspension viscosity; *on-state*, which is characterised by the excess shear stress; and *power requirement*, which is characterised by the current density. This article discusses the potential benefits of using ER fluids and outlines some of the basic design problems. Some simplified expressions are derived that permit parametric design of ER fluid-based systems.

POTENTIAL BENEFITS AND BASIC PROBLEMS IN DESIGN WITH ER FLUIDS

ELECTRO-RHEOLOGICAL (ER) fluids possess the ability to change from a liquid to a solid when subjected to an electric field and back again, to a fluid, when the field is removed. Typically they consist of micron-sized particles suspended in an electrically insulating continuous phase. The reversible change occurs in milliseconds and the particles are generally believed to structure under the influence of the electric field. The principal induced structural property is conventionally termed the *yield* or *excess shear stress*, although the more colloquial term, *strength of the fluid*, is often used. Put simply, these terms define the ability of the electrically stressed fluid to withstand applied external forces.

Because ER fluids enable direct control of movement by electrical means, a major design consideration is to select devices and applications that are either impractical or impossible when using conventional means. The consequent advantages are:

- electronic control of mechanical devices in greatly simplified systems with fewer or no moving parts
- loose tolerance, and hence inexpensive, mechanical/hydraulic systems
- quiet systems, as the mechanical parts do not move
- rapid reacting systems more akin to the electronics that control the device
- electronically controlled infinitely variable torque transmission or braking
- multi-aspect performance, such as simultaneous position and velocity control of a mechanism
- electronically compensated changes in performance due to wear in other components

- no heavy copper windings and heavy steel parts
- no wear, and hence long life expectancy
- the introduction of active characteristics into otherwise passive devices

However, there is a need to recognise that ER fluids have their own particular design characteristics and it is not simply a case of removing an existing component and replacing it with an equivalent ER device. The ER device needs to be developed as a fully integrated package; that is, a complete systems capability is required. This will vary with the particular application, but essentially capability is required in electronics, control, rheology, mechanical engineering, and knowledge of manufacturing. Additional capabilities may also be required for specific devices, e.g., vibration expertise for mounting systems.

An ER device has five basic elements (as indicated in Figure 1), which, shown in block diagram, form a common arrangement of:

1. Mechanical components
2. ER fluid
3. High voltage source
4. Controlling logic circuit
5. Feedback elements to provide control signals

These five elements need to be successfully integrated to produce a practical ER device. Two other components are the load to be controlled, item 6, and any additional sensors, item 7. The load, shown as a mass in Figure 1, is taken to be any controlled variable, e.g., position, velocity, or acceleration. Other aspects in the design procedure to be considered include the electrical conductivity, power requirement, thermal conductivity, and internal temperature rise—particularly under shear, viscosity, and response time.

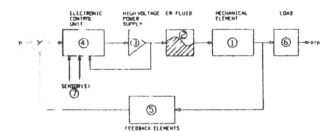

Figure 1. *Block diagram of an ER system.*

Preliminary design considerations of an ER fluid device are relatively straightforward. However, detailed design considerations to optimise the device are complex. In all but the simplest device, a large amount of design data needs to be assembled. For many of the above five elements, selection and compromise in the design method are well documented and organised. For example, in mechanical design, materials selection for different applications is common; for control, numerous sensors are available with different characteristics; and with electronics, parametric design methods are readily available. However, when we consider the fluid itself, parametric design data is either unavailable or incomplete. It is to the clarification of this limitation that the article is directed.

As ER fluids offer a degree of activity in an otherwise passive system, a common method used to make an initial appraisal of an ER fluid is the difference between the forces transmitted with and without the field applied. The ratio has been termed the "control ratio". However, such an appraisal is by no means straightforward. By way of example, let us consider a simple concentric cylinder clutch arrangement. The no-field transmitted torque is a function of the shear rate and the fluid viscosity, assuming that the ER fluid is essentially Newtonian. As an initial approximation, the with-field transmitted torque can be taken as independent of the shear rate. Consequently, as the shear rate is increased, the no-field torque increases and the control ratio falls. In reality, the with-field torque is not independent of the shear rate and as a consequence, control ratio is a poor predictor of the device performance. The case is even more complex for a stabilised ER fluid. Here, the no-field behaviour can be highly non-Newtonian and hence the no-field transmitted torque also varies rapidly with the shear rate.

Similar complex evaluations are needed when assessing the power requirement. Current tends to fall as shear rate is increased, but is compensated to some extent by a rise in the temperature of the fluid due to viscous heating. The final steady-state value of current will therefore depend upon the heat transfer characteristics of the fluid. Implicit in this statement is the assumption that the thermal conductivity is independent of the ER effect; an assumption that is, in all probability, false.

However, despite these difficulties one must start somewhere. The route chosen here is to identify the significant variables and concentrate on those that affect the mechanical and electrical design of the device. The ER fluid is integrated into this process by a consideration of its properties. Specifically, we are interested in the excess shear stress, the effect of both volume fraction and continuous phase viscosity, and the power requirements of the device.

As an introductory step, it can be assumed that the off-state behaviour is effectively characterised by the viscosity; the on-state behaviour is characterised by the excess shear stress; and finally, the power required to energise the device is characterised by the current density. These factors will be discussed in turn and are relevant to the behaviour of numerous ER fluids in static, Couette, or Poiseuille flow. The dispersed phases included in this discussion are silicas, ceramics, polymers, salts of polymers, and polyacine radicals. The continuous phases covered are mainly paraffins or silicone oils, although a number of other materials have been studied, such as hydrocarbons, fluorocarbons, esters, aliphatic oils, and fluorosilicone oils. The observations noted are similar in all cases.

OFF-STATE BEHAVIOUR

The viscosity of an ER fluid is a function of the shear rate, the temperature of the fluid, the viscosity of the continuous phase, μ_f, and the nature and volume fraction of the dispersed phase.

Effect of Shear Rate

Because they are concentrated suspensions, ER fluids are intrinsically non-Newtonian materials. However, within the normal range of usage—that is, less than 40% volume fraction—the fluid behaves approximately as a Newtonian fluid—shear stress is proportional to the shear rate. A small degree of shear thinning is often observed, typically a 10% reduction in viscosity from 100 to 1000 sec^{-1}, and this should be noted. Additives used to modify the physical stability, e.g., a surfactant or other methods of introducing interparticle forces, will exaggerate the non-Newtonian properties of the suspension. For example, a typical rheological stabilising agent may introduce a gelled structure that breaks down under shear, but is rapidly reformed when the shear is removed. Such materials may exhibit a small yield stress or alternatively a very high viscosity at low shear rates. This needs to be considered both in the development of devices and from a handling point of view. Fluids with even a small yield stress or a high viscosity do not flow easily.

Effect of Temperature

The effect of temperature on the suspension viscosity is dominated by the temperature dependence of the continuous phase. The conventional relationship between viscosity and the temperature is observed with ER fluids made from both paraffin and silicone oil. The low viscosity index of silicone oil is well known and is observed in ER fluid suspensions manufactured using these materials. This relationship is independent of volume fraction, although the magnitude of the viscosity will vary.

Effect of Volume Fraction and the Continuous Phase

It is necessary to consider the effect of both the continuous phase and the volume fraction of the dispersed phase. Silicone oil is available in a number of variants with viscosities from a few mPas to tens of thousands of mPas. Indicated in the inset of Figure 2 is the variation of the suspension viscosity for different volume fractions of dispersed phase when suspended in two oils. All data were taken at a shear rate of 500 sec^{-1} and at a temperature of 30°C. The silicone oil was a 10 cSt fluid. The paraffin, Cereclor 50 LV, was approximately seven times as viscous as the silicone oil and hence, the resulting ER fluid was also more viscous. To overcome such differences in assessing the viscosity of suspensions, it is common practice to divide the viscosity at a given volume fraction by the continuous phase viscosity. The resulting term is called the relative viscosity, μ_r. This is shown as a function of volume fraction in Figure 2. As evident from the figure, these fall approximately on top of each other and clearly indicate the value of this concept in determining the viscosity of ER fluids. From this graph, the viscosity of any lithium polymethacrylate-based ER fluid can be evaluated. A 30% volume fraction suspension has a viscosity approximately 4.2 times the oil viscosity. So, for the 8.8 mPas (10 cSt) silicone oil, the ER fluid viscosity is approximately 37 mPas. However, for a silicone oil of 460 mPas (500 cSt), the viscosity of the ER fluid will be approximately 1900 mPas.

By using Figure 2, the viscosity of ER fluids can be quickly calculated. The range of viscosities for an ER fluid can be large, from 4 mPas to 700,000 mPas, depending upon the volume fraction and the particular oil chosen. Previous research has indicated that the effect on the excess shear

stress due to the continuous phase viscosity for silicone oils is minimal. This, of course, does not imply that the total shear stress is independent of the continuous phase viscosity. The total shear stress is the sum of the excess shear stress and the shear stress due to the shear rate, the Newtonian component.

The variation of viscosity with volume fraction in a number of concentrated suspensions has been studied in great depth over many years, although no real agreement has been reached as to the precise relationship between the viscosity and volume fraction. Most relationships are semiempirical in nature and the viscosities are related via the maximum packing factor of the system being studied. A series of papers by Quemada [1977, 1978(a), 1978(b)] details one of the more simple relationships, which is of the form:

$$\mu_r = (1 - \tfrac{1}{2}k\phi)^{-2} \qquad (1)$$

where $k = 2/\phi_m$, ϕ_m is the maximum packing fraction. This varies from 0.74 for face centred cubic or close packed, to 0.525 for simple cubic packing. These values assume mono-dispersed spherical particles. ϕ is the volume fraction. Fitting the data from Figure 2 to the above equation returns a value for k of 3.661, giving ϕ_m as 0.55. This compares favourably with the experimental value for the maximum packing fraction of 0.58 obtained via high speed centrifugation of samples.

Therefore, from the simple expression:

$$\mu_s = (1 - 1.83\phi)^{-2} * \mu_f \qquad (2)$$

the suspension viscosity can be calculated for any oil at any volume fraction of dispersed phase.

ON-STATE BEHAVIOUR

The on-state characteristics of the fluid depend upon two primary factors, the applied field and the volume fraction, and two secondary factors, the shear rate and the fluid temperature. Only the primary factors are discussed here.

Effect of the Electric Field

There is little conjecture over the fact that as the field strength is increased, the excess shear stress also increases. It is commonly assumed in simulation work that the structure formed under the applied field is due to polarisation forces. These are often taken as dipole-dipole or enhanced dipole-dipole forces (Klingenberg et al., 1989). Consequently, an E^2 dependence is predicted. However, when data from fluids is studied, such behaviour is not that apparent. In Figure 3, the excess shear stress for an ER fluid in Poiseuille flow is plotted against field strength. Two sets of data are shown, one up to fields of 3.0 kV/mm, and the other up to fields of 6.5 kV/mm. If an expression of the form

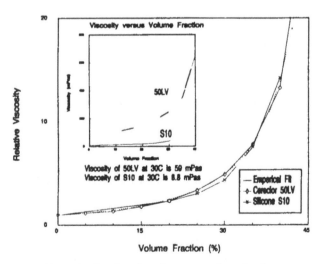

Figure 2. Relative viscosity versus volume fraction.

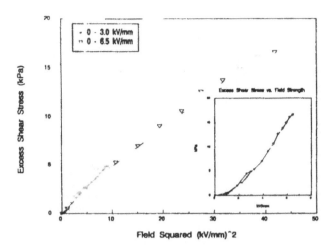

Figure 3. *Excess shear strength versus field squared.*

$A_1 * E^2$ is fitted to the data, the coefficient A is found to be 0.533 for the 3.0 kV/mm data set, and 0.431 for the 6.5 kV/mm data set. Furthermore, the 6.5 kV/mm data set demonstrates significant curvature. This is despite the overlap of the raw data as illustrated in the inset of Figure 3. A simple square relationship is unable to cover both sets of data. For example, the excess shear stress at 5.92 kV/mm, taking the coefficient from the 3.0 kV/mm data set, results in a stress of 18.68 kPa, whereas the measured value was only 14.92 kPa. However, in the interests of simplicity, such an expression can be used, provided care is exercised and an appropriate value for the coefficient is used.

At this stage, it is not really appropriate to speculate on the optimum expression for the field strength dependence. Further work is required in this area. Data from other materials together with estimates of the errors involved are necessary. This is especially important as previous work has indicated that confidence limits can be large and effectively mask any distinctive form for the curve. However, at this stage, an E^2 is appropriate *if* used with care.

Effect of Volume Fraction

As the volume fraction is increased at fixed conditions of field strength and shear rate, the excess shear stress also increases. Figure 4 shows a typical plot for a pH 9.0 lithium salt of polymethacrylic acid, 15.5% water content, in Couette flow at 30°C and at a field strength of 2.35 kV/mm. Two continuous phases are indicated, again a 10 cSt silicone oil and Cereclor 50 LV.

Similar plots can be generated at other field strengths, shear rates, and temperatures although, of course, the magnitudes will vary. Also shown in Figure 4 is the data for the relative viscosity, μ_r, modified to be $\mu_r - 1$. From this it is quickly apparent that the viscosity rises faster than the excess shear stress. Consequently, the difference between the on-state and off-state cannot be achieved simply by increasing the volume fraction of the dispersed phase.

Experimental work that does exist on the variation of excess shear stress, usually called yield stress in the literature, is related to gels or other materials with yield stress on the order of a few pascals. Data and experimental prediction on materials with yield stress on the order of kPa are rare. Surprisingly, wood pulps (Bennington et al., 1990) often result in high yield stress suspensions. The variation of yield stress with volume fraction in such materials is usually characterised by an expression of the form:

$$T_e = A_2 * \phi^n \tag{3}$$

where n lies between 2 and 4. Fitting the experimental data to the above expression results in:

$$T_e = 43.85 * \phi^{2.517} \tag{4}$$

with a regression coefficient of 0.99611. The expression can be relaxed slightly by fixing n to 2.5 to give:

$$T_e = 43.20 * \phi^{2.5} \tag{5}$$

with a regression coefficient of 0.99610, an insignificant difference.

It would appear that the excess shear stress is dependent on the volume fraction raised to approximately 2.5 or to the power of 5/2.

POWER REQUIREMENTS

The power requirement of an ER device can be characterised by the product of the fluid volume, the field strength, and the current density (Bares and Carlson, 1989). Given that J and E are related in some fashion, the power requirement is effectively controlled by the current density. The usual objective is to minimise this. The current density is a

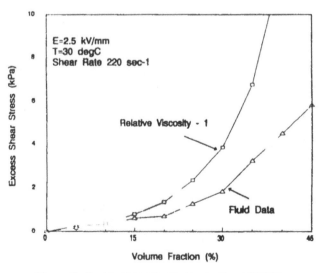

Figure 4. *Excess shear stress versus volume fraction.*

strong function of both the field strength and the fluid temperature, and a weak function of both the shear rate and the volume fraction.

Effect of the Electric Field

Since the power requirement of any ER device is governed by the current density and the field strength, surprisingly little attention has been focused on the characteristic form and, by implication, the origin of the current density. What publications there have been are generally vague, alluding to current density being either proportional to the field (Bullough et al., 1987) raised to the power n, with n varying from less than 1 to 5, or exponential with the field (Conrad et al., 1989). The question that occurs is: Are any of these approximations appropriate? If not, are there more soundly based governing expressions?

A typical current density plot is indicated in Figure 5, along with an exponential fit. It is apparent that such a fit is reasonably appropriate. In fact, an exponential expression is the limiting case of the Butler-Volmer equation (Bockeris and Reddy, 1977) for interfacial charge transfer. The other limit is a linear one at low electric fields. Hence, from a consideration of the charge-transfer reaction mechanism at the interfaces, some of the confusion regarding the current flow in ER fluids can be resolved. If the formulation of the ER fluid is such that the operating field results in the charge transfer mechanism operating in the low field region, a linear relationship will be observed. Correspondingly, when operated at the high field limit, an exponential relationship will be observed. As the Butler-Volmer equation is the difference between two exponential terms, this can be driven to a simplified solution of the form:

$$J = \alpha \sinh (\beta E) \qquad (6)$$

where α is given by $A_3 \exp^{(-\epsilon/kT)}$ and β is given by $\delta q/kT$.

Here, A_3 is a constant, ϵ is an activation energy, k is

Boltzmann's constant, T is the temperature, δ is a jump distance, and q is the charge carried. The sinh expression not only describes the form of the conductivity curve extremely well, but is also based on well-founded theoretical principles. However, as ER fluids are used almost without exception at the high field limit, the simpler exponential-type expression can often be used in the development of devices.

Effect of Temperature

It has been observed that increasing the temperature of an ER fluid will result in an increased current requirement. This is a direct result of the temperature appearing as the denominator of the exponential term, $\exp^{(-\epsilon/kT)}$, of Equation (6). It is also observed that when the log of the current density is plotted against the electric field data, a straight line results. This is directly related to the high field limit as described previously. This now leads to the possibility of values being ascribed to the unknowns in the expression $J = \alpha \sinh (\beta E)$ from measurements of the current density and electric field. From such measurements, at different temperatures, it is possible to arrive at values for A_3, the proportionality constant, ϵ, the activation energy, and δ, the jump distance. Typical values for A are on the order of $1 * 10^{15}$; ϵ, between 0.6 to 1.0 eV; and for δ, approximately $2 * 10^8$ m or 200 Å. It is tempting to ascribe A_3 to the number of charges available for the conduction process and ϵ as a characteristic activation energy to force the charge across the potential barrier defined by the jump distance, δ. Superficially, one might expect the jump distance to be the interparticle distance. However, the magnitude is such as to suggest that it is probably associated with the surface characteristics of the particle.

Effect of Shear Rate and Volume Fraction

One advantage of the rate limited diffusion process described above is that it enables the observed variations in current density with volume fraction and shear rate to be explained. First, with an increased volume fraction, we are simply increasing the number of charge carriers. One would therefore expect a clear relationship between the A_3 term and the volume fraction. This would appear to be the case, as both α and β are seen to increase with increased volume fraction. An increase in α implies an increase in the number of conducting species. The data suggests values on the order of 10^8 at 5% volume fraction rising to 10^{15} at 40% volume fraction. The increase in β adds weight to the speculation that it is not the interparticle distance, as the latter should fall dramatically as the volume fraction is increased. The increase in β suggests an increase in the jump distance δ, which is a little difficult to explain. At this stage, it is not possible to adequately explain the variations of current density with shear rate. Because such explanations would, at this stage, be only speculative, they are not covered in this article.

This is necessarily only a brief summary of the current

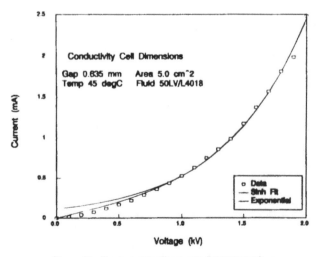

Figure 5. *Current density versus field strength.*

flow in an ER fluid. Further data will be presented elsewhere (Brooks, 1993). It is hoped, however, that sufficient data and explanation have been given to enable some of the apparent mystery to be removed from this issue. Current flow in ER fluids is related to the charge transfer mechanism occurring at the particle-oil interface. For practical purposes, this can be reduced to a simple exponential expression of the form:

$$J = A_4 \exp(-\beta E) \qquad (7)$$

SUMMARY OF EXPRESSIONS

An attempt has been made to reduce the volume of data associated with ER fluids to a more manageable level. Simplified expressions have been put forward and the limitations in their usage noted. Typical values for the coefficients associated with lithium polymethacrylate-based ER fluids have also been indicated.

In the off-state, the behaviour is controlled by the viscosity, which can be expressed as:

$$\mu_s = (1 - 1.83\phi)^{-2} * \mu_f \qquad (2)$$

This is a general expression and enables the suspension viscosity to be assessed for different continuous phases and different volume fractions.

The on-state is characterised by the excess shear stress. This is a function of both the electric field and the volume fraction. The simplest relationship for the electric field is:

$$T_e = A_1 * E^2 \qquad (8)$$

However, care is needed as the coefficient is often inadequately defined. Typical values for A_1 in both Couette and Poiseuille flow are between 0.4 and 0.6.

Volume fraction changes can be characterised by:

$$T_e = A_2 * \phi^{5/2} \qquad (3)$$

A typical value for A_2 in both Couette and Poiseuille flow is 43.

The power requirements are governed by the current density. This can be expressed in the form of:

$$J = \alpha \sinh (\beta E) \qquad (6)$$

where $\alpha = A_3 \exp^{(-\epsilon/kT)}$ and $\beta = \delta q/kT$.

Typical values for A_3 are on the order of $1 * 10^{15}$, ϵ is on the order of 0.6 to 1.0 eV, and δ is on the order of $2 * 10^{-8}$ m.

In practice, an exponential expression of the form:

$$J = A_4 \exp(-\beta E) \qquad (7)$$

is often adequate.

REFERENCES

Bares, J. E. and J. D. Carlson. 1989. "Electrorheological Fluid Design: An Overview", *Prod. 2nd Int. Conf. on ER Fluids*. Lancaster, PA: Technomic Publishing Co., Inc.

Bennington, C. P. J., R. J. Kerekes and J. R. Grace. 1990. "The Yield Stress of Fibre Suspensions", *Cand. J. of Chem. Eng.*, 68:748-757.

Bockeris, J. O'M., and A. K. N. Reddy. 1977. *Modern Electrochemistry* 2. Rosetta, NY: Plenum, pp. 862-878.

Brooks, D. A. To be published. "Conduction Processes in Electro-Rheological Fluids".

Bullough, W. A., D. J. Peel and R. Firoozian. 1987. "Electrical Characteristics and Other Considerations for Electro-Rheological Fluids", *Prod. 1st Int. Conf. on ER Fluids*, North Carolina State University, Engineering Publications.

Conrad, H., Y. Chen and A. F. Sprecter. 1989. "Electrorheology of Suspensions of Zeolite Particles in Silicone Oil", *Prod. 2nd Int. Conf. on ER Fluids*. Lancaster, PA: Technomic Publishing Co., Inc.

Klingenberg, D. J., F. van Swol and C. F. Zukoski. 1989. "Dynamic Simulation of Electrorheological Suspensions", *J. Chem. Phys.*, 91(12):7888-7895.

Quemada, D. 1977. "Rheology of Concentrated Disperse Systems and Minimum Energy Dissipation Principle. I. Viscosity-Concentration Relationship", *Rheo. Acta*, 16:82-94.

Quemada, D. 1978(a). "Rheology of Concentrated Disperse Systems. II. A Model for Non-Newtonian Shear Viscosity in Steady Flow", *Rheo. Acta*, 17:632-642.

Quemada, D. 1978(b). "Rheology of Concentrated Disperse Systems. III. General Features of the Proposed Non-Newtonian Model", *Rheo. Acta*, 17:642-653.

Optical Effects of Electro-Rheological Fluids

L. W. Hunter,* F. F. Mark, D. A. Kitchin, M. R. Feinstein,
N. A. Blum, B. R. Platte and F. G. Arcella
The Johns Hopkins University
Applied Physics Laboratory
Johns Hopkins Road
Laurel, MD 20723-6099

D. R. Kuespert and M. D. Donohue
The Johns Hopkins University
Chemical Engineering Department
Baltimore, MD 21218

ABSTRACT: Measurements have been made of the 525 nm region (green) light transmission through suspensions of silica in a liquid fluorocarbon. Transmission dropped when electrostatic field strengths up to 1350 V/mm were applied perpendicular to the light path, but no drop was detected for fields parallel to the light path. The transmission drop did not depend on incident polarization. We observed that an ER fluid changes the polarization of transmitted light with and without an applied electric field, but we were unable to resolve the dependence on the electric field.

INTRODUCTION

ELECTRO-RHEOLOGICAL (ER) fluids are suspensions of polarizable particles in dielectric fluids. When ER fluids are subjected to an electric field, the particles in the fluid form fibrils—long chains that often span from electrode to electrode when there is no flow to break them up. The fibrils change the rheological properties of the suspension.

Willis Winslow (1950) found that a concentrated suspension of silica in mineral oil exhibited a dramatic increase in apparent viscosity when subjected to a strong electric field; specifically, Winslow's fluid acquired elastic Bingham fluid behavior. In the following years, the number of ER fluid recipes has increased dramatically and many papers have been devoted to their properties and applications such as electric clutches, shock absorbers, valves, and hydraulic control systems. A few of the useful review articles and recent conference proceedings include Gast and Zukoski (1989), the *Proceedings of the Blacksburg Conference on Recent Advances in Adaptive and Sensory Materials and their Applications* (1992), and the *Proceedings of the Carbondale Electrorheological Fluids Conference* (1991).

Here we describe experiments to investigate interactions between ER fluids and light, with a view toward applications to variable optic filters and displays. In particular, we sought to observe whether application of an electric field to an ER fluid changes the light transmission, and whether the change in transmission depends on polarization. We also sought to observe any change in polarization as a result of passage through the ER fluid, with and without an applied electric field. These effects have been predicted by Park (1988) and by Adriani and Gast (1988, 1989). In other work, Feinstein (1991) has used classical electromagnetic theory to predict the absorption and scattering of microwave radiation by small nonspherical weakly conducting particles in a nonconductive medium; the theory shows that for incident signals of a given wavelength, there is a particle size and orientation that maximizes absorption. Measurements of turbidity have been reported by Jordan, Wong and Shaw (1988) and Jordan and Shaw (1989). Polarization-dependent optical measurements on silica-based ER fluids were presented by Smith and Fuller (1989). In addition, optical transmission measurements were presented by Ginder and Elie (1991) in independent work.

ER FLUID RECIPES

Our carrier liquids are 3M Company "Fluorinert" brand fluorocarbon dielectric liquids, namely "FC-87" and "FC-70", of which FC-87 is lighter, less viscous and more volatile. Diatomaceous earth (SiO_2) particles were equilibrated with atmospheric humidity, pretreated in the FC-87 until the liquid evaporated, then added to FC-70 to a concentration of about 0.9% by weight, then decanted twice. Our estimate of final particle concentration is about 0.1% by weight. The concentration remained constant for the times required for this experiment. Particle settling appeared to be minimal for at least 30 minutes as long as an electric field was not applied; the electric field tended to pull particles toward the electrodes.

We also experimented with suspensions of nanoparticles

---

*Author to whom correspondence should be addressed.

in FC-70. The particles, fabricated by Professor J. L. Katz (JHU/ChemEng), are solid (not hollow) and have a measured surface area of 130 m²/g. The diameter is estimated to be 20 to 60 nm. Concentrations of 0.14 and 0.28% by weight were used. The suspension contained flocculent gelatinous cloud-shaped bodies that agglomerated with time.

EXPERIMENTAL APPARATUS

Figure 1 is a schematic of the apparatus.

We used a DC power supply with variable voltage (rated to 10 kV, with a 10 mA maximum). A switch/resistor bridge box provided voltage control, and allowed measurements to be made by a low voltage digital voltmeter. The voltage could be set at the supply and switched to the cells via safety wiring. An electrically insulated, light-tight cover for the spectrometer was built. Maximum voltages were about 4–5 kV, giving a highest electrostatic field strength of about 1350 V/mm.

A Perkin-Elmer Model 330 spectrophotometer was used to measure transmission of light through the ER fluid. It is a differential instrument with the highest transmission path acting as reference. Reference and sample optical paths were equal with the reference side being blank (air).

Two kinds of spectrometer cells were fabricated for measuring changes in the intensity of transmitted light. In one, the light is parallel to the applied electric field and hence to the ER particle "strings" that form; in the other, the light is perpendicular to the field and the strings. In both cases, standard acrylic spectrophotometer cuvettes were modified to accept the ER fluid and associated electrodes.

Figure 2 illustrates the conceptual details.

For the configuration where the light beam is parallel to the strings so that the light beam passes through electrode plates, our first design used electrodes made with 30 nm of chromium sputtered onto thin lexan strips; these electrodes provided 5% transmission equally at all wavelengths (neutral density), but the chromium layer was not robust. In the cell assembly, thicker lexan strips were inserted into the cuvettes first, next to the four inner surfaces, so as to decrease the electrode spacing to 3.12 mm and keep the field

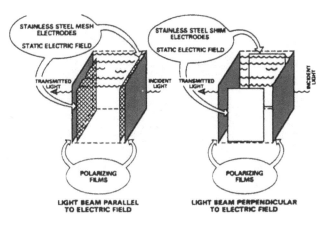

Figure 2. *Conceptual details of the experiment.*

strength in a convenient range. High voltage leads were attached and potting compound was used to secure the structure.

Other optically-transmitting electrode coatings are available, including cadmium-tin-oxide (Tu et al., 1991), indium-tin-oxide, and transparent conducting polymers.

The chromium electrodes eventually deteriorated, probably due to electrolytic effects or other chemical action. In addition, the electrode spacing was found to be too small for convenient control of the ER fluids. Another set of transparent electrodes was fabricated from 0.05 inch mesh stainless steel screen attached to thin Plexiglas sheets. The spacing was 4.8 mm for this cell.

For the cell where the light beam is perpendicular to the fibrils, the light does not pass through the electrodes. In our first design, the electrodes were made of chromium-gold plated lexan strips about 3 mm apart. A subsequent improved design used stainless steel shim stock electrodes spaced 5.0 mm apart.

To investigate light polarization effects, combinations of horizontal and vertical polarizing filters (HN-22 type) were used as polarizers and analyzers (in both reference and sample path beams). These polarizer sheets do not polarize well in the blue, restricting operation to green, yellow and red. It was estimated that the spectrophotometer light beam is 70–80% horizontally polarized.

RESULTS

Our qualitative observations were as follows. When the field is applied to the 0.1% diatomaceous earth recipe, some fibrils were visible to the naked eye. In addition, the particles move back and forth between electrodes at a speed that increases with the field strength; ultimately, they tend to collect at the electrodes (with field still on) so that the fluid gradually clears. When the field is turned off, the particles begin to redisperse, but this takes time.

Application of an electric field to the 0.14% nanoparticle

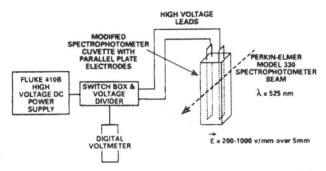

Figure 1. *Apparatus schematic (electric field perpendicular to light beam).*

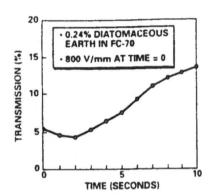

Figure 3. *Transmission of light perpendicular to electric field.*

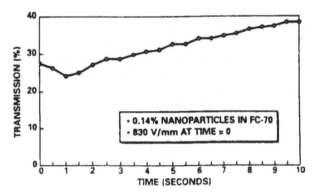

Figure 4. *Transmission of light perpendicular to electric field.*

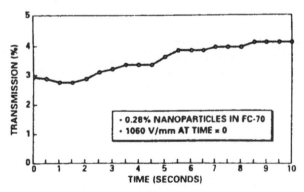

Figure 5. *Transmission of light perpendicular to electric field.*

fluid caused the particles to migrate to the electrodes, where they formed fine threads that did not span the gap between the electrodes. Particles in the 0.28% nanoparticle fluid also collected at the electrodes but formed clumps rather than threads. Thus, the nanoparticle suspensions, like the diatomaceous earth suspensions, gradually clear when an electric field is applied.

Transmission measurements with 525 nm (green) light are presented in Figures 3–5. With the light propagating perpendicular to the (DC) electric field, we observed an initial drop in transmission. Subsequently, the field caused a gradual and continuing increase in transmission. The increase occurred in the parallel orientation as well.

The diatomaceous earth recipe gave an initial drop of 10 to 30%, perpendicular. In the cell with light parallel to the electric field, no initial transmission drop was observed.

Transmission through the 0.14% nanoparticle fluid was less than that through the 0.1% diatomaceous earth fluid. Transmission through the 0.28% nanoparticle fluid was about comparable to the 0.1% diatomaceous earth fluid.

The transmission perpendicular to the field through the 0.14% nanoparticle recipe dropped 11% initially. The transmission parallel to the field showed no initial drop (starting at 19% and rising to 22% when the run ended).

The transmission perpendicular to the field through the 0.28% nanoparticle recipe dropped 5% initially in the one case we ran. The transmission parallel to the field showed no initial drop.

The initial transmission drop that we observe occurs within the first second after the field is turned on, which is the limit of time resolution of the spectrophotometer used. The subsequent recovery occurs within 10 seconds as clearing continues.

For the diatomaceous earth recipe, we varied field strength from 200 to 1060 V/mm but found no correlation with the transmission drop.

We did not observe any change in the relative transmission drop when the polarization of the incident light beam was varied.

To assess induced polarization, we placed crossed filters—one upstream and one downstream of the cell. The crossed filters, by themselves, reduce the transmission to 0.4% of the maximum possible transmission, independent of wavelength. When an ER cell containing only carrier liquid is placed between the two crossed filters, the transmission rises slightly. With diatomaceous earth ER fluid present, the transmission rises again to about 2%. When the electric field is applied perpendicular to the light, the transmission drops initially, by 10–30%, just as it does without filters. When the field is applied parallel to the light path, the transmission does not drop initially. In both orientations, the gradual clearing is observed.

Thus, unaligned particles change the polarization of the transmitted light; however, we were unable to resolve whether there is any additional effect on the polarization when the particles were aligned by imposition of an electric field. The relative transmission drop does not seem to be connected to polarization effects.

DISCUSSION

We attribute the initial drop in transmission to microscopic fibril formation and the subsequent rise to particle rearrangement and collection at the electrodes. This interpretation is consistent with Klingenberg and Zukoski (1990).

The experimental results provide ample opportunity for further work. Investigation of the effects of surfactants on ER fluid suspensions would provide a tool for stabilizing ER fluid formulations. Production of monodisperse particles for ER suspensions remains a technical obstacle to this investigation. The use of nanoparticles in ER fluid investigations should be examined further, since a nanoparticle suspension would provide less equipment abrasion and handling difficulties than traditional microparticle suspensions.

CONCLUSIONS

Experiments with visible light detected a reduction in transmission when an electric field is applied perpendicular to the light propagation direction, but not when the field is parallel. The transmission drop did not depend on incident polarization. We observed that an ER fluid changes the polarization of transmitted light with and without an applied electric field, but we were unable to resolve the dependence on the electric field.

ACKNOWLEDGEMENTS

The authors thank Professor John P. Coulter (Lehigh University) for beneficial discussions and for providing numerous references, and Professor Joseph L. Katz (Johns Hopkins University, Chemical Engineering Department) for providing SiO_2 nanoparticles. The reference by Smith and Fuller was drawn to our attention by one of the referees.

This work was supported by IR&D funds.

REFERENCES

Adriani, P. M. and A. P. Gast. 1988. "A Microscopic Model of Electrorheology", *Phys. Fluids*, 31(10):2757–2768.

Adriani, P. M. and A. M. Gast. 1989. "Predictions of Birefringence and Dichroism of Hard Sphere Suspensions in Combined Electric and Shear Fields", *Journal of Chemical Physics*, 91(10):6282–6289.

Feinstein, M. R. 1991. "Scattering and Absorption by Particles in an Electro-Rheological Fluid", JHU/APL Report FIB-2-91U-012.

Ginder, J. and D. Elie. 1991. "Dielectric and Optical Probes of Electrorheological Fluid Structure", presentation to *Electrorheological Fluids Conference, Carbondale, October 15–16*.

Gast, A. M. and C. F. Zukoski. 1989. *Adv. Coll. Int. Sci.*, 30:153.

Jordan, T. C., W. Wong and M. T. Shaw. 1988. "A Rheo-Optical and Materials Approach to Electrorheology", Annual Report, *Conference on Electrical Insulation and Dielectric Phenomena, October 16–20*, pp. 493–499.

Jordan, T. C. and M. T. Shaw. 1989. "Electrorheology", *IEEE Trans. Electrical Insulation*, 24(5):849–878.

Klingenberg, D. J. and C. F. Zukoski. 1990. "Studies on the Steady-Shear Behavior of Electrorheological Suspensions", *Langmuir*, 6:15–24.

Park, O. O. 1988. "Electrohydrodynamics of Rigid Macromolecules with Permanent and Induced Dipole Moments", *Journal of Rheology*, 32(5):511–531.

Smith and Fuller. 1989. Presentation to *First International Conference on ER Fluids*, H. Conrad, A. F. Sprecher and J. D. Carlson, eds., Raleigh: North Carolina State University.

Tu, L., F. Schubert, H. M. O'Bryan, Y. Wang, B. E. Weir, G. J. Zydzak and A. Y. Cho. 1991. *Applied Physics Letters*, 58(8):790.

Winslow, W. M. 1950. "Field Controlled Hydraulic Device", U.S. Patent 2,661,596.

1992. *Recent Advances in Adaptive and Sensory Materials and Their Applications*. Proceedings of the Conference held at Virginia Polytechnic Institute and State University, Blacksburg, April 27–29, 1992. Lancaster, PA: Technomic Publishing Company, Inc.

Proceedings of Electrorheological Fluid Conference, Southern Illinois University, Carbondale, 15–16 October, 1991. To be published by World Scientific Publishing.

Properties of Zeolite- and Cornstarch-Based Electrorheological Fluids at High Shear Strain Rates

A. KOLLIAS AND A. D. DIMAROGONAS*
Department of Mechanical Engineering
Washington University
One Brookings Dr., Campus Box 1185
St. Louis, MO 63130

ABSTRACT: The mechanical and electrical properties of zeolite- and cornstarch-based ER fluids are explored using a concentric cylinder viscometer. With cylinder radius to gap size ratio on the order of 400, we achieved shear strain rates ranging from 3000 to 20,000 sec⁻¹. Two different volume fractions of zeolite and cornstarch suspensions in silicone oil were tested at a constant temperature of 30°C. The shear stresses and electric resistivity of the fluids were computed from experimental measurements as functions of shear strain rate.

Although the cornstarch ER fluids indicated anomalous behavior, the zeolite ER fluids were found to be more stable under high shear strain rates and electric fields and were characterized by a Bingham viscoplastic model-like behavior. Dynamic yield stress was increased in proportion to a power of the electric field. The higher the volume fraction of dispersions, the higher the yield stress. Plastic viscosity, as defined by the Bingham model, remained unchanged with the variation of electric field for a fixed volume fraction.

This work was motivated by the need for applications to rotor-bearing control, where for usual bearings, the shear rate is much higher than the maxima of most of the published experimental results. Moreover, for high shear rates, the validity of the Bingham model was questioned by many authors without specific experimental justification, which is attempted here for two different kinds of dispersed particles.

INTRODUCTION

ELECTRORHEOLOGICAL (ER) fluids consist of dielectric particles (mainly hydrophilic) dispersed in electrically nonconducting liquids. Externally applied voltage rapidly transforms these fluids from Newtonian to viscoplastic materials with higher apparent viscosity.

Klass and Martinek (1967) reported experiments with silica-based ER fluids for shear strain rates ranging from 0 up to 2300 sec⁻¹. Their data show that a threshold value of shear stress is required to initiate the fluid flow at zero shear strain rate and high electric field, the "static yield stress". During shearing, an excess shear stress (above the shear stress at zero electric field), which is generally different than the static yield stress, needs to be maintained. This excess shear stress has been termed by Kraynik (1989) and Stangroom (1989) as "dynamic yield stress" since it refers to dynamic shearing where the hydrodynamic forces dominate. Constant dynamic yield stress through high shear strain rates implies viscoplastic behavior described by the Bingham model.

Among the ER fluids reported, those with organic dispersions are characterized by a much larger static than dynamic yield stress according to:

1. Bullough (1989), who reported tests for a 28% volume fraction of starch in transformer oil and for shear strain rates in the range of 0 to 350 sec⁻¹
2. Seed, Hobson and Tozer (1989), who reported tests for a 42% weight fraction of cornstarch in silicone oil and for shear strain rates from 0 to 18,000 sec⁻¹

Both reports described anomalous behavior at low shear strain rates, and dynamic yield stress decreasing with shearing.

The zeolite particles as electrorheological dispersions in a nonelectrically conducting oil were patented by Filisko and Armstrong (1988) and experiments were reported for shear strain rates from zero up to 470 sec⁻¹. The moisture effect was investigated by Filisko and Radzilowski (1990), who reported that zeolite suspensions exhibit an ER effect even in small amounts of water. An extensive study was reported by Conrad, Chen and Sprecher (1989) where the electrical and mechanical properties of three types of zeolites were investigated for small shear strain rates (up to 8.5 sec⁻¹). A more recent work for zeolite ER fluids by Conrad, Chen and Sprecher (1991) concentrates on the shear yield stress dependence on shear strain and shear strain rate (from 0 up to 500 sec⁻¹).

Viscometry of ER fluids at low shear rates has dominated the literature of electrorheology in recent years because of the relatively high ER effect (static yield stress) in this range of shear. The low shear rate ER effect (or the static yield stress) has been credited for the periodic breaking and reforming of the fibrous structure in the fluid (Klingenberg and Zukoski, 1990; Conrad, Chen and Sprecher, 1991).

---

*Author to whom correspondence should be addressed

The ER effect at high shear rates for many ER fluids is questioned by Block and Kelly (1988) since the fibrils are destroyed by the hydrodynamic forces and the dielectric properties of the ER fluids are affected by the shear strain rate. The polarization mechanism between the dispersed particles is generally affected by the rate of shearing, but the behavior of a specific ER fluid depends on its particular ER mechanisms.

Since there are applications (such as journal bearings) where the shear rate is very high, the aim of this work is to test the validity of the Bingham fluid model at high shear rates, particularly for ER fluids based on zeolite and cornstarch dispersions in silicone oil.

EXPERIMENTAL SET-UP

Apparatus

For high shear strain rate measurements, a concentric cylinder viscometer was built (Figure 1) driven by a 3 hp dc motor with a speed controller. The driving speed was measured directly on the rotating cylinder with a magnetic pickup.

One inner brass cylinder (bob) with a 100 mm diameter and a 102 mm length was used. Two outer stationary aluminum cylinders (cups) with inner diameters of 100.5 mm and 101.4 mm and of equal length yielded two radial gap sizes of 0.25 mm and 0.7 mm (or 0.5% and 1.4% of the bob radius). The rotating cylinder was concentric with respect to the aluminum cup but shorter in length, leaving a 4.7 mm gap at the bottom and 3.0 mm at the top.

The inner cylinder was driven by the motor and the outer cylinder was held stationary with a steel rod. The torque on the rod was recorded with a strain gage. The experimental error of the torque measurements, including both the electric circuit of the strain gage and the rod linkage, did not exceed 0.5%.

The bottom gap between the inner cylinder and the bottom of the outer cylinder was filled with fluid, while at the top the fluid formed a free surface. Since the bottom was electrically insulated, the fluid was sheared under zero electric field. The torque contributed by the bottom part can be computed from $T_b = \pi\eta\omega R_1^4/2H$ [parallel disc shearing (Bird, Armstrong and Hassager, 1987)], where η is the zero electric field Newtonian viscosity of the fluid, ω is the rotational speed of the inner cylinder of radius R_1, and H is the gap at the bottom of the viscometer. The contribution of the shearing at the flat bottom of the viscometer was estimated for the 27% zeolite at shear rate 20,000 sec^{-1} and shear stress 2000 Pa to be lower than 0.6% of the total measured torque T.

Vibration analysis of the rotor, including the brass cylinder, yielded a first natural frequency of 3120 rpm, which is way above the maximum driving speed of 1300 rpm used during the experiments.

Voltage Source

The electric field was produced by a commercial dc power supply (BG Electronics, model P231) and amplifier (PTK Electronics, 0–3.5 kVdc, with voltage regulator), delivering up to 3500 Vdc and 1 mA. The positive electrode was the inner cylinder for all the experiments.

Measurements of the actual voltage in the electrodes were made just after the amplifier and before the viscometer (Figure 1). The electric current going through the fluid was recorded with an ammeter in series with the viscometer. For insulation of the metallic parts, Plexiglass was used, as shown in Figure 1, with a dielectric constant of approximately 3.5.

Materials

The dispersed particles and the carrier oil used are both commercially available. The zeolite Z3125 (sodium aluminosilicate) was supplied by Sigma Chemicals Co. with the following characteristics (Filisko and Armstrong, 1988):

1. Chemical formula $Na_{86}[(AlO_2)_{86}(SiO_2)_{106}] \times H_2O$
2. Nominal pore diameter 9–10 Å
3. Particle diameter < 10 microns
4. Water capacity approximately 34% by weight

The cornstarch is a common commercial powder, but exact technical information was not available to the authors. The particle size is estimated to be 3 times larger than that of zeolite.

For the carrier nonconducting fluid, the 200 silicone fluid (polydimethylsiloxane), 50 cSt, provided by Dow Corning Co., was used. Water and other exempt compounds, according to the manufacturer, do not exceed 9 g/l. The actual kinematic viscosity was found to be 47 cSt at 30°C and the density was 975 kg/m³ (dynamic viscosity 0.0458 Pa sec). In the tests, pure Newtonian behavior characterized the silicone oil under shearing and high electric fields. The dielectric constant was about 2.5 (Conrad, Chen and Sprecher, 1989).

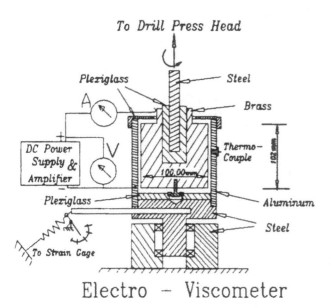

Figure 1. *The electroviscometer apparatus*

In this work, two mixtures for each dispersion were prepared. Table 1 shows the physical properties of the mixtures. The high speed testing provided sufficient stirring to the dispersions for a homogeneous mixture. The cornstarch and silicone fluid were not dried and were used as received. An ER fluid based on nondryed zeolite yielded dielectric breakdown in very low electric fields. For this reason, the zeolite was heated for 6 hours at 180°C, losing 2.5% of its initial weight. The exact amount of the remaining water in the zeolite was not determined.

Temperature Measurement and Control

Although concentric cylinder viscometers appear to have many design advantages, they produce considerable amounts of viscous heat during the operation at high shear rates. In the experiments, the temperature was controlled by forced air cooling and the run was terminated when the temperature exceeded a pre-set limit (30.8°C).

The temperature was maintained at 30 ± 1°C, even at maximum shear rates and electric fields. The error introduced by the ± 1°C difference did not affect the viscosity measurements more than 2% according to sensitivity tests conducted in low electric fields.

To monitor the temperature, a thermocouple was attached to the outer cylinder in contact with the fluid. The maximum temperature variation across the gap is given by Schlichting (1979) as $\delta T = \eta(\dot{\gamma}_m h)^2/8k$, where $\dot{\gamma}_m$ is the maximum shear rate, h is the gap width, η is the viscosity and k is the thermal conductivity of the fluid. A conservative estimate of δT can be found using the silicone oil thermal conductivity ($k = 0.155$ W/m°C at 50°C, supplied by Dow Corning Co.). The estimated δT is 1.2°C for 27% zeolite (at $\dot{\gamma} = 15,000$ sec^{-1}, $\eta = 0.105$ Pa sec) and 3.1°C for the 32% cornstarch (at $\dot{\gamma} = 8000$ sec^{-1}, $\eta = 0.125$ Pa sec). The temperature variation across the gap is assumed to be small enough to yield substantial viscosity differences.

This temperature variation also affects the yield stress across the gap. Conrad, Chen and Sprecher (1989) found that for zeolite-based ER fluids, the yield stress for very low shear rates ranging from 0.08 to 8 sec^{-1} increases proportionally to the exp(− 1/temperature) with a maximum around 100°C (the vaporization temperature of water). Consequently, for a temperature variation of 2°C, a difference of 3.5% in the yield stress must be expected.

Table 1. Data and physical properties of the mixtures used.

Suspension Used	Volume Fraction (%)	Density (kg/m³)	Viscosity at Zero Volts (Pa sec)	Diameter of Aluminum Cup Used (m)
Zeolite	20	1089	0.065	0.1005
Zeolite	27	1130	0.105	0.1005
Cornstarch	20	998	0.115	0.1014
Cornstarch	32	1018	0.125	0.1014

EXPERIMENTS

Procedure

Before each start-up, the inner cylinder was rotated at high speeds to stir and warm the mixture up to 30°C. When the desired temperature was reached, voltage was set at a fixed value. At low speed of the cylinder rotation, the voltage was turned on and measurements were taken while increasing the speed at discrete steps. When the temperature exceeded the preset limit, the test was interrupted for cooling. A complete series of measurements over the speed range (under the same voltage) usually required one interruption at low electric fields, and two at high electric fields.

The current going through the fluid was monitored during the experiments.

Observations

ZEOLITE ER FLUIDS

The gap width used, $h = 0.25$ mm, was equivalent to at least 25 particle diameters, enough for a homogeneous mixture.

Voltage and current were recorded separately from the torque measurements as functions of the mean shear rate $\omega R_1/h$, where ω is the rotational speed of the inner cylinder with radius R_1 and length L. The behavior of the mean electric field intensity E (voltage/h) is given in Figures 2a and 2b for the 20% and 27% zeolite ER fluids. Using the viscometer area $S = 2\pi R_1 L$, the current density J (current/S) was plotted in Figures 3a and 3b. While the inner cylinder was stationary, local dielectric breakdown of the fluid occurred (largest current at zero shear rate). This caused the measured electric field to be well under the initial setting (at high shear rates) and the electric current to increase. A slight motion of the cylinder made the electric field nearly approach the setting. Further acceleration gradually stabilized the electric field and the current density to the high shear rate values. Higher shear rates were needed at higher electric fields to reach steady conditions. Only torque measurements which correspond to the steady behavior of the electric field (right of the dashed lines in Figures 2a and 2b) were further analyzed.

The fibrils formed by the particles were probably used as bridges for the electric charges to pass from one electrode to the other. Thus, when a fibrillated structure existed between the electrodes, current transmitted through the fluid. The fact that a current passed through even at very high shear rates implies that particle coalitions were sufficient to transmit electric charges.

It is also apparent from Figures 3a and 3b that the higher the electric field, the higher the current going through the fluids. The product EJ gives the electric power consumption per unit volume of the sheared fluid, and it was constant for a fixed voltage setting. The supplied electric power was not enough to keep the electric field constant due to shear rate dependent electric resistivity of the fluids. The electric resistivity (equal to E/J) of the 20% and 27% zeolite ER fluids is plotted in Figures 4a and 4b. It is apparent that the

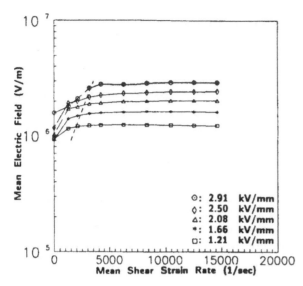

Figure 2a. *Mean electric field vs. mean shear strain rate for the 20% zeolite per volume in silicone oil.*

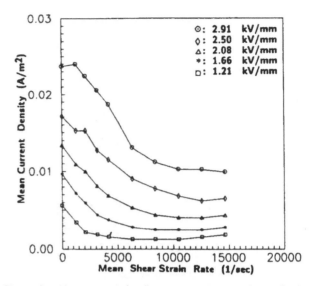

Figure 3a. *Mean current density vs. mean shear strain rate for the 20% zeolite per volume in silicone oil.*

higher the shear rate, the higher the resistivity, but also, the higher the electric field, the lower the resistivity.

Torque measured during a steady electric field was linearly proportional to the rotational speed and thus a Bingham fluid behavior was implied. Measurements with voltage applied before acceleration and those with voltage applied after acceleration, at the same shear rate, were identical and quite reproducible.

The fluid, during high speed shearing, responded instantly (in order of 0.01 sec) to the application of the electric field.

CORNSTARCH ER FLUIDS

The gap width used was $h = 0.7$ mm. At zero and low speeds, local dielectric breakdown was observed occasionally with no extended duration.

In the start-up of the rotor, under high electric field, a "stick-slip" effect was observed with the outer cylinder oscillating and the torque readings fluctuating. The "stick-slip" effect is common in friction of solids, where the friction coefficient is a decreasing function of the velocity and converges to a steady value for very high velocities. The similarity between friction and electrorheological yield stress is obvious. The "stick-slip" effect corresponds to the transition from the static to dynamic yield stress with increasing shear rate, and it is apparent that the ER material was at a solid state during "stick-slip" (see also Figure 8).

Further acceleration resulted in lower torque, followed by a short range of speeds with constant torque. At higher speeds, the torque increased proportionally for the 20%, but not for the 32% cornstarch ER fluid.

At high shear rates, the measurements were not quite

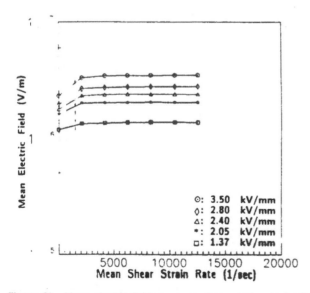

Figure 2b. *Mean electric field vs. mean shear strain rate for the 27% zeolite per volume in silicone oil.*

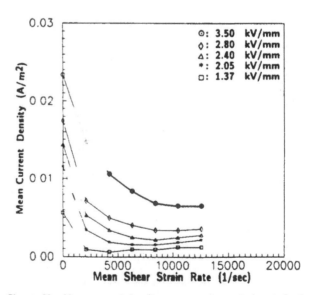

Figure 3b. *Mean current density vs. mean shear strain rate for the 27% zeolite per volume in silicone oil.*

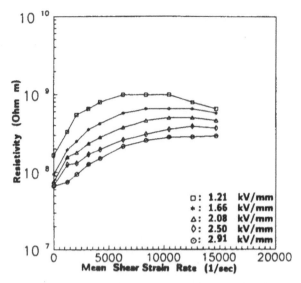

Figure 4a. Electric resistivity vs. mean shear strain rate for the 20% zeolite per volume in silicone oil.

reproducible and differences occurred for the same speed and voltage, probably due to the "stick-slip" effect.

Accelerating and decelerating the inner cylinder gave a small hysteresis to the torque-speed results.

Similar observations for the cornstarch ER fluid have been made by Bullough (1989).

Limitations to the maximum speed were imposed by the inertia forces, which made the flow unstable and the torque not increase with speed. This is a common effect in cylindrical viscometers and is not related specifically to the ER effect. Testing with both cylindrical cups showed that flow instability occurs at high velocities and not necessarily at high shear strain rates.

With ϱ as the density and μ as the viscosity of the ER fluids ($\mu = \eta$ at zero electric field), the maximum Reynold's number, $Re = \varrho \omega R_1 h / \mu$, was 20.95 for the zeolite

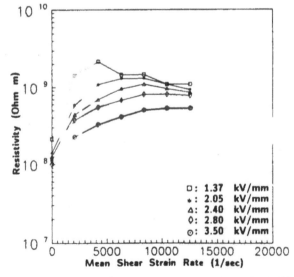

Figure 4b. Electric resistivity vs. mean shear strain rate for the 27% zeolite per volume in silicone oil.

and 34.1 for the cornstarch experiments. The Taylor number, $Re(h/R_1)^{1/2}$, is used to predict initiation of flow vortices in Couette viscometers (Schlichting, 1979). With a suggested upper limit value of 41.3 to avoid vortices, the Taylor number was 1.46 for the zeolite ER fluids and 4.03 for the cornstarch ER fluids. Therefore, the fluid flow in both types of dispersions and gaps was laminar.

EXPERIMENT ANALYSIS AND RESULTS

At high speeds, the ER fluids tested (except for the 32% cornstarch mixture) behaved like Bingham fluids, since a linear relationship between torque and speed was observed. The Bingham fluid model in one-dimensional shear flow may be written as

$$\tau = \tau_v + \tau_o = \mu \dot{\gamma} + \tau_o \tag{1}$$

where μ is the "plastic" viscosity, τ_o is the dynamic yield shear stress and τ_v is the viscous shear stress. In the case of ER fluids, μ is not necessarily the viscosity of the zero electric field η and represents the slope of the stress-shear rate lines at high shear rates. The dynamic yield stress, τ_o, can be found by extrapolation of the above stress-shear rate line down to zero shear rate and is a constant for a specific electric field. If the viscosity is not constant but varies with shear rate, then either the Bingham fluid model cannot be used or the dynamic yield stress should also be considered as a function of shear rate. An interesting discussion for the Bingham model of the ER fluids and Couette type viscometers is given by Stangroom (1989).

In the tests, zeolite suspensions produced linear torque-speed relation (Bingham fluid behavior), and the viscosity at higher electric fields μ was nearly the same as η for zero electric field. The cornstarch suspensions did not behave as Bingham-like fluids at small shear rates. In fact, at high shear rates, the torque-speed relation was linear for the 20% cornstarch mixture with different slopes for different electric fields.

Using the Reiner Equation to fit the experimental data for Couette flow of Bingham materials (Oka, 1960), the viscosity is given by:

$$\mu = (\tau_1 - \tau_2 - 2\tau_o \ln s) 2\omega \tag{2}$$

where ω is rotational speed of inner cylinder, τ_1 is the shear stress on the inner moving cylinder wall ($r = R_1$), τ_2 is the shear stress on the inner wall of the cup ($r = R_2$) and $s = R_2/R_1$. The shear strain rate at $r = R_2$ is given by Equation (1):

$$\dot{\gamma} = (\tau_2 - \tau_o) \mu \tag{3}$$

The shear stress is given as a function of the torque T as $\tau = T/2\pi L r^2$, where L is the height of the inner cylinder. The ratio of the stresses at $r = R_1$ (inner cylinder) and $r = R_2$ (inner radius of cup) is on the order of $\tau_1 \tau_2 = s^2 = (R_2 R_1)^2 = 1.01$ for zeolite and 1.028 for cornstarch experiments. Therefore, the variation of the stress across the gap is quite small and, with negligible error, the dynamic

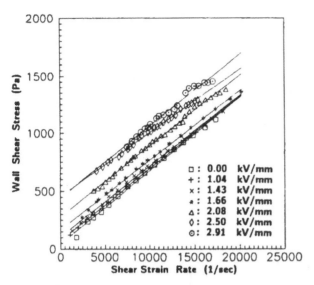

Figure 5. *Shear stress vs. shear strain rate at the inner wall of the cup for the 20% zeolite per volume in silicone oil.*

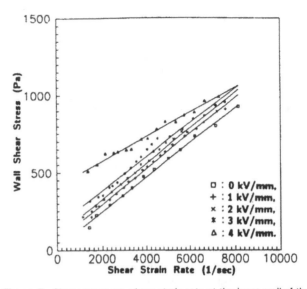

Figure 7. *Shear stress vs. shear strain rate at the inner wall of the cup for the 20% cornstarch per volume in silicone oil.*

yield stress at zero speed can be computed from Equation (2), setting $\omega = 0$ and $\tau_2 = \tau_o$ (Oka, 1960):

$$\tau_o = \tau_{10}(1 - 1/s^2)/2 \ln s \qquad (4)$$

where $\tau_{10} = T_o/2\pi LR_1^2$ and T_o is the dynamic yield torque computed from the extrapolation of the linear fit of torque vs. speed experimental data down to the zero speed. In brief, the computed τ_o from Equation (4) is used with Equation (2) to yield the viscosity μ and the shear strain rate at the inner wall of the cup from Equation (3).

The region of very low speeds, where the ER fluid is forming a plug in the inner surface of the cup, cannot be investigated for the gaps used in this work, not only because the difference $\tau_1 - \tau_2$ is very small, but also because of the electrical problems encountered at this range of shear rates

(see "Observations" section). A larger gap is required for low shear rate studies and plug monitoring.

Using the above relations, the cup inner wall shear stress, τ_2, and the shear strain rate were computed for the ER fluids (Table 1) and for a variety of electric fields. Since the 32% cornstarch ER fluid yielded a nonlinear relationship between torque and speed, the mean shear stress, $\tau = T/2\pi LR^2$, was computed as a function of the mean shear rate $\omega R/h$, where $R = (R_1 + R_2)/2$. Results are shown in Figures 5, 6, 7 and 8. A linear fit (Bingham model) of the computed points is drawn wherever it is applicable.

The electric field intensity at the cup inner wall can be computed from $E_2 = V/[R_2 \ln s]$. The electric field intensity in Figures 5–8 represents an average voltage gradient E ($= V/h$) and it differs from E_2: 0.25% for the 0.25 mm gap and 0.7% for the 0.7 mm gap. The maximum electric field

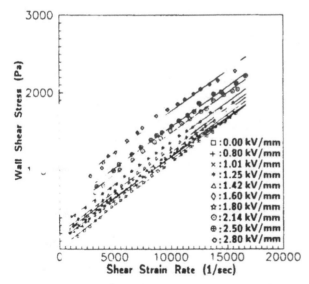

Figure 6. *Shear stress vs. shear strain rate at the inner wall of the cup for the 27% zeolite per volume in silicone oil.*

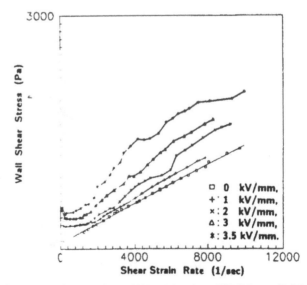

Figure 8. *Shear stress vs. shear strain rate at the inner wall of the cup for the 32% cornstarch per volume in silicone oil.*

shown in each of Figures 5–8 is very close to the limit of the dielectric breakdown.

In Figures 5 and 6 (zeolite ER fluids), a parallel shift of the stress-shear rate fit lines (Bingham model) to a higher stress level was observed as the electric field was increased. The slope of the fit lines (viscosity) was not affected by the higher electric field, although an increased scatter of the data can be observed and is probably due to the interruptions for cooling. The heat generation at higher electric fields is larger because the apparent viscosity is increased.

For the 20% cornstarch fluid (Figure 7), the Bingham model gives the best fit. While the viscosity at low electric fields appears unchanged, it tends to decrease for higher electric fields; the ER effect (and dynamic yield stress) decreases with the shear rate for very high electric fields. In Figure 8, the behavior of the 32% cornstarch fluid is more complicated. It is obvious that the stress level is shifted by increasing the electric field, but at high shear rates, the behavior was peculiar because the viscosity increased. This behavior has not been explained yet.

The dynamic yield stress is given as a function of the mean electric field, E, in Figure 9 for the zeolite and Figure 10 for the cornstarch ER fluids. In Figure 9, two sets of calculated data are presented, corresponding to 20% and 27% volume fractions of zeolite. For a fixed electric field, the higher the volume fraction, the higher the yield stress. In fact, the relation between static yield stress and volume fraction of the zeolite is linear, as has been reported by Conrad, Chen and Sprecher (1991).

In Figure 9, two fits for each volume fraction data set were plotted. The solid curve corresponds to an exponential fit and the dashed curve to a power fit of the data points. The formulas used were $\tau_o = AE^p$ for the power and $\tau_o = B \exp(qE)$ for the exponential fit. For E in V/m:

- 20% zeolite:
  $A = 5.32 \times 10^{-12}$ Pa $(m/V)^p$, $p = 2.17$
  $B = 17.63$ Pa/exp(V/m), $q = 1.308 \times 10^{-6}$

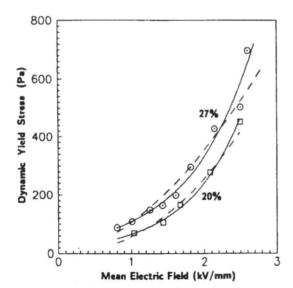

Figure 9. Dynamic yield stress vs. mean electric field for the 20% and 27% zeolite per volume in silicone oil.

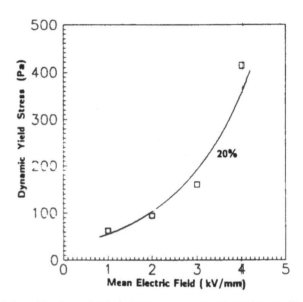

Figure 10. Dynamic yield stress vs. mean electric field for the 20% cornstarch per volume in silicone oil.

- 27% zeolite:
  $A = 3.53 \times 10^{-9}$ Pa $(m/V)^p$, $p = 1.74$
  $B = 35.21$ Pa/exp(V/m), $q = 1.122 \times 10^{-6}$

Similarly, for the 20% cornstarch fluid (Figure 10), only an exponential fit was obtained with $B = 29.70$ Pa/exp(V/m), $q = 6.211 \times 10^{-7}$.

DISCUSSION AND CONCLUSIONS

The temperature control is an important factor at high shear rates because it affects both viscosity and yield stress. Keeping the error due to temperature variations to a low level, the high shear strain rate and electric field properties were investigated.

The cornstarch ER fluids appeared to have, to some extent, unpredictable behavior. At low shear rates, they exhibit solid-like friction with a "stick-slip" phenomenon and a much higher static than dynamic yield stress. Low speed devices (brakes, clutches) are possible applications for these ER fluids.

Zeolite suspensions proved to be Bingham-like fluids if exposed to strong electric fields. Their response to the electric field was fast and at high shear rates, easily reproducible. The zeolite ER fluids performed quite well at high shear rates. The electric power consumption given by $P = EJhS$ was relatively small for the ER fluids used. For example, the electric power for the 20% zeolite at high shear rates and $E = 2.5$ kV/mm, $J = 0.007$ A/m² was $P = 0.13$ W.

The Bingham model describes adequately the high shear rate behavior of the zeolite ER fluids. An extension of this model to low shear rates seems quite possible according to the results reported by Filisko and Armstrong (1988) and Conrad, Chen and Sprecher (1991).

However, the current density was found to be much higher in the low shear rates than in the high shear rates. According to Conrad, Chen and Sprecher (1989), the yield stress of zeolites in low shear rates is proportional to $J^{1/5}$. This is an indication for the existence of a transient region from the static to dynamic yield stress.

LIST OF SYMBOLS

E = electric field intensity
h = radial gap width
H = bottom gap height
J = electric current density
k = thermal conductivity
R = mean radius $(R_1 + R_2)/2$
R_1 = inner cylinder radius
R_2 = outer cup inner radius
Re = Reynold's number
T = torque
V = voltage
$\dot{\gamma}$ = shear strain rate
δT = temperature difference
η = zero electric field viscosity
μ = slope of the stress-shear rate line, viscosity
ϱ = fluid density
τ_o = dynamic yield stress
τ_1 = shear stress at the inner cylinder wall
τ_2 = shear stress at the inner wall of the cup
ω = rotational speed of inner cylinder

ACKNOWLEDGEMENT

The authors gratefully acknowledge the financial support provided by Showa Wire and Cable Co.

REFERENCES

Bird, R. B., R. C. Armstrong and O. Hassager. 1987. *Dynamics of Polymeric Liquids–Fluid Mechanics, Vol. 1, 2nd Ed.* New York: Wiley-Interscience, p. 37.

Block, H. and J. P. Kelly. 1988. "Electro-Rheology", *Journal of Physics D: Applied Physics*, 21(12):1661–1677.

Bullough, W. A. 1989. "Miscellaneous Electro-Rheological Phenomena, Part II", in *Proc. 2nd Int. Conf. on ER-Fluids*, J. D. Carlson, A. F. Sprecher and H. Conrad, eds., Lancaster, PA: Technomic Publ. Co., Inc., p. 124.

Conrad, H., Y. Chen and A. F. Sprecher. 1989. "Electrorheology of Suspensions of Zeolite Particles in Silicone Oil", in *Proc. 2nd Int. Conf. on ER-Fluids*, J. D. Carlson, A. F. Sprecher and H. Conrad, eds., Lancaster, PA: Technomic Publ. Co., Inc., p. 252.

Conrad, H., Y. Chen and A. F. Sprecher. 1991. "The Strength of Electrorheological (ER) Fluids", in *Proc. of the Int. Conf. on ER-Fluids*, R. Tao, ed., World Scientific Publ. Co., p. 195.

Filisko, F. E. and W. E. Armstrong. 1988. "Electric Field Dependent Fluids", U.S. patent 4,744,914.

Filisko, F. E. and L. H. Radzilowski. 1990. "An Intrinsic Mechanism for the Activity of Alumino-Silicate Based Electrorheological Materials", *Journal of Rheology*, 34(4):539–552.

Klass, D. L. and T. W. Martinek. 1967. "Electroviscous Fluids I. Rheological Properties, II. Electrical Properties", *Journal of Applied Physics*, 38(1):67–80.

Klingenberg, D. J. and C. F. Zukoski. 1990. "Studies on the Steady-Shear Behavior of Electrorheological Suspensions", *Langmuir*, 6(1):15–24.

Kraynik, A. K. 1989. "Comments on ER-Fluid Rheology", in *Proc. 2nd Int. Conf. on ER-Fluids*, J. D. Carlson, A. F. Sprecher and H. Conrad, eds., Lancaster, PA: Technomic Publ. Co., Inc., p. 445.

Oka, S. 1960. "The Principles of Rheometry", in *Rheology–Theory and Applications, Vol. 3*, F. R. Eirich, ed., New York: Academic Press, p. 33.

Schlichting, H. 1979. *Boundary-Layer Theory, 7th Ed.*, New York: McGraw-Hill: temperature variation, pp. 277–279; Taylor number, pp. 525–527.

Seed, M., G. S. Hobson and R. C. Tozer. 1989. "Macroscopic Behaviour of ER-Fluid", in *Proc. 2nd Int. Conf. on ER-Fluids*, J. D. Carlson, A. F. Sprecher and H. Conrad, eds., Lancaster, PA: Technomic Publ. Co., Inc. p. 214.

Stangroom, J. E. 1989. "The Bingham Plastic Model of ER-Fluids and Its Implications", in *Proc. 2nd Int. Conf. on ER-Fluids*, J. D. Carlson, A. F. Sprecher and H. Conrad, eds., Lancaster, PA: Technomic Publ. Co., Inc., p. 199.

ER Fluid Applications to Vibration Control Devices and an Adaptive Neural-Net Controller

SHIN MORISHITA*
Department of Naval Architecture and Ocean Engineering
Yokohama National University
156 Tokiwadai, Hodogaya-ku, Yokohama 240, Japan

TAMAKI URA
Institute of Industrial Science
The University of Tokyo
7-22-1 Roppongi, Minato-ku, Tokyo 106, Japan

ABSTRACT: This article describes several applications of electro-rheological (ER) fluid to vibration control actuators and an adaptive neural-net control system suitable for the controller of ER actuators. ER fluid is a kind of colloidal suspension, and has a notable characteristic feature in that its apparent viscosity can be controlled in response to applied electric field strength. Viscosity can be varied in a wide range and the response time is very short, as short as several milliseconds. According to previous studies, one promising application is a controllable damper. In the present article, four applications are proposed: a shock absorber system for automobiles, a squeeze film damper bearing for rotational machines, a dynamic damper for multi-degree-of-freedom structures and a vibration isolator. Furthermore, an adaptive neural-net control system, composed of a forward model network for structural identification and a controller network, was introduced for the control system of these ER actuators. As an example study of intelligent vibration control systems, an experiment was conducted in which the ER dynamic damper was attached to a beam structure and controlled by the present neural-net controller so that the vibration in several modes of the beam was reduced with a single dynamic damper.

INTRODUCTION

ELECTROVISCOUS effect, which indicates the reversible change in apparent viscosity of a fluid subjected to an electric field, was first reported by Duff in 1896. Though various studies from several aspects have been undertaken to date (Bjornstahl and Snellman, 1937; Andrade and Dodd, 1946), this unique effect has not generated much industrial interest, since these studies were focused primarily on the mechanism of the effect itself. Moreover, in these studies, the variations in apparent viscosity were limited to several tenths of a percent—a not so attractive range for industrial applications.

The electro-rheological (ER) effect is considered to be part of an electroviscous effect and is defined for colloidal suspensions. In 1949, Winslow reported remarkable viscosity variation when an electric field was applied to certain colloidal suspensions. He also suggested several industrial applications in his paper. Attractive characteristic features of ER fluid are the wide variation of its apparent viscosity and its quick response time.

There have been various attempts to apply ER fluid mechanical components: ER damper (Stevens, Sproston and Stanway, 1984), ER clutch (Stevens, Sproston and Stanway,

1988), ER engine mount (Duclos, 1987; Petek, Goudie and Boyle, 1988; Ushijima, Takano and Noguchi, 1988), and ER valve (Bullough and Stringer, 1973). However, taking into account not only the welcome characteristics of ER fluid described above, but also the undesirable imperfections, such as sedimentation of dispersed particles or loss of its effect under high shear rate, one promising application is a controllable damper. One of the authors has been attempting to apply ER fluid to mechanical components: a shock absorber system (Morishita and Mitsui, 1991), a squeeze film damper bearing (Morishita and Mitsui, 1992a), a damping-controllable dynamic damper (Morishita and Kuroda, 1991), and a controllable vibration isolator (Morishita and Mitsui, 1992b).

In the present article, four application studies investigated by the present authors are presented. The controllable shock absorber system proved to be effective for the improvement of drivability of automobiles. High speed flexible rotors could pass through critical speeds with the controllable squeeze film damper quite smoothly by setting the most appropriate supporting damping for each vibration mode of the rotor. The damping-controllable dynamic damper could reduce the vibration response of a structure in several modes with a single dynamic damper. Stiffness, as well as damping capacity, could be varied by the ER vibration isolator. In addition to these applications, an adaptive neural-net control system is proposed in this article. Because ER fluid

*Author to whom correspondence should be addressed

properties are nonlinear functions of temperature, shear rate or electric field strength, it is difficult to develop a prevalent controller. Thus, a neural-net controller was chosen as a suitable controller for ER devices. The performance of the controller was studied with the damping-controllable dynamic damper in numerical simulations of a two-degree-of-freedom structure and experiments on a beam structure.

TYPICAL PROPERTIES OF ER FLUID

An ER fluid consists of a suspension of fine semiconducting particles in a dielectric liquid (Klass and Martinek, 1967; Block and Kelly, 1988; Jordan and Shaw, 1989), and the apparent viscosity of ER fluid can be varied by the applied electric field strength. Various kinds of semiconducting particles, such as cornstarch grain, cellulose, silica-gel, alumina powder, ion exchange resin or other artificial materials, have been shown to be applicable to ER fluid (Sugimoto and Kondo, 1976; Inoue, 1989). The particle size ranges from several tenths of a nanometer to 10 micron meter. Subjected to an electic field, many chains of particles dispersed in the ER fluid are formed between the electrodes, and these chains of particles cause the resistance to flow (flow mode), or resistance to shear movement of the electrodes (shear mode) as shown in Figure 1. This increase of resistance is closely associated with the increase of viscosity, and the rheological behavior resembles that of Bingham plastic fluid.

The typical properties of the ER fluid are shown in Figure 2. The present properties are based on the experimental results obtained by a rotational type viscometer developed by one of the authors. Because an ER fluid presents Bingham plastic behavior, the fluid has a certain degree of yield stress at zero shear field. The yield stress increases as the field strength is increased, and the shear stress of the present ER fluid is shown under various electric field strengths. The gradient of each line indicates the Newtonian viscosity of the fluid. Relative shear stress is defined as the ratio of shear stress with an electric field to that without an electric field. It should be noted that the viscosity variation is reversible, and that the response time needed for the variation is as short as several milliseconds (Morishita and Mitsui, 1991).

ER FLUID APPLICATIONS

Shock Absorber

One of the promising applications of ER fluid to mechanical components is a controllable damper. Several

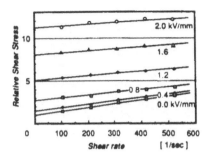

Figure 2. Typical properties of ER fluid.

authors have already proposed controllable shock absorber systems for automobiles using ER dampers (Petek, 1992). The author also designed a prototype ER damper and experimental studies were performed.

The construction of the present ER shock absorber system is shown schematically in Figure 3. In order to apply the ER fluid to a shock absorber system, electrodes were set up in a pathway of ER fluid. In this regard, two specific types of construction have been proposed. In one type, the electrodes are equipped in a piston so that the electrodes themselves may reciprocate in the ER fluid along a cylinder. In the other type, the electrodes are installed in a bypass attached additionally to the cylinder. In the present experiment, the first type was chosen for simpler construction. The positive and negative electrodes were placed alternately and fixed to the piston rod through an insulator.

Figure 4 shows one of the experimental results. It shows the amplitude of the damping force at various piston speeds and electric field strengths. Although the amplitude of the damping force depended on the piston speed, it could be controlled up to 2.5 times as large in the absence of electric field in the low piston speed range. According to numerical simulations, this capacity is sufficient enough to improve the drivability of automobiles (Morishita and Mitsui, 1991).

Squeeze Film Damper

The second application is a controllable damper for rotating machines. Because a flexible rotor may be operated over several critical speeds in a high speed rotating machine, a squeeze film damper is often set up for the stabilization of the rotor. Though it is well known that optimum supporting damping capacities for each critical mode of the flexible rotor exist, the damping capacity of conventional type squeeze film damper is governed by the clearance of the squeeze film and the viscosity of lubricant. By controlling the supporting damping capacity, the optimum damping can always be given to the rotor.

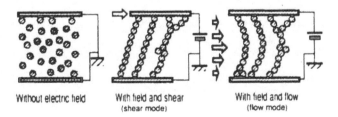

Figure 1. Mechanism of ER effect.

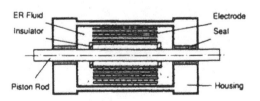

Figure 3. ER shock absorber.

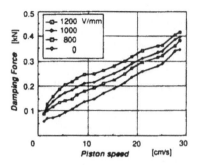

Figure 4. Damping properties.

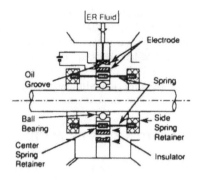

Figure 6. ER squeeze film damper.

A flexible rotor system with an ER squeeze film damper is depicted schematically in Figure 5. A flexible shaft with three fly-wheels was supported by squeeze film dampers at both ends of the shaft. For simplicity, the ER fluid was applied to only one of the squeeze film dampers and the other was fed with normal lubricant. The shaft was designed to be flexible enough such that at least the first and the second vibration modes were within the operating speed. The designed critical speeds without any damping were 1452 rpm for the first mode and 3254 rpm for the second.

Detailed construction of the ER squeeze film damper is shown in Figure 6. For the application of ER fluid to the squeeze film damper, a pair of electrodes were set up. In the present case, each electrode corresponded to the inner and outer cylinder of the damper; one electrode was set up in the damper housing, and the other was attached to the center spring retainer independently. Each cylinder was 10 mm in width including oil groove. Insulators (ceramics) were placed between the electrode and the housing, and also between the electrode and the center spring retainer. The ball bearing with center spring retainer was supported by several centering pin springs. These springs were supported by the side spring retainers at both sides, and the side spring retainers were adjustable in the vertical direction with an adjuster. The clearance between electrodes was 0.16 mm and gravity fed with ER fluid from a circumferential groove.

In the present experiment, the damping ratios corresponding to vibration modes were evaluated at various electric field strengths by the free vibration response to an impact to the rotor. Figure 7 shows the variation of the spectral plots at various electric field strengths. The spectrum had a rather round peak at about 1400 cpm in the absence of electric field, but the peak tapered gradually and practically vanished at the field strength of 1.2 kV/mm, which indicated

that the optimum damping was given to the rotor. As the field strength was increased further, another peak, comparatively steep, appeared and grew gradually.

The first and second critical speeds of the rotor and the equivalent damping ratios are shown in Figure 8. With the increase in field strength, the second critical speed went up from about 3800 rpm to 5000 rpm, followed by the first critical speed elevation from 1400 rpm to 1900 rpm. During this elevation of the first mode, two identical critical speeds appeared. At the field strength where the critical speeds went up, the damping ratio had an extreme value. In the same way, the ER squeeze film damper can eliminate several critical speeds with a single damper by controlling its damping capacity (Morishita and Mitsui, 1992a).

Dynamic Damper

The third application is a damping-controllable dynamic damper. A dynamic damper is considered to be one of the most reliable passive-type vibration actuators. One dynamic damper can decrease the amplitude at a single eigenfrequency of the structure. In the present application study, the author tried to reduce the amplitude in several vibration modes of the structure with a single damping-controllable dynamic damper, in which damping capacity was controlled by an adaptive neural-net controller. The construction of the neural-net controller will be shown in the following section.

The ER dynamic damper construction is shown in Figure

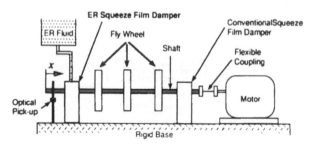

Figure 5. Flexible rotor system.

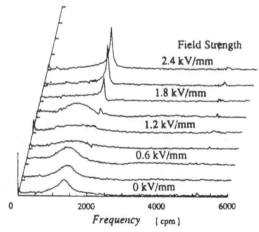

Figure 7. Power spectrum (response to an impact).

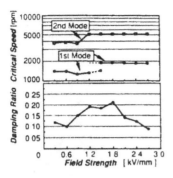

Figure 8. *Critical speed and equivalent damping ratio.*

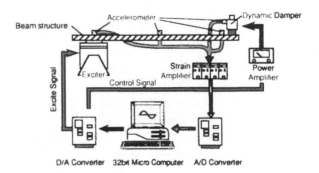

Figure 10. *Experimental setup for dynamic damper.*

9. The present ER dynamic damper was essentially composed of two concentric cylinders and two coil springs. The inner and outer cylinders were made of mainly copper and used as electrodes. A reserve tank of the ER fluid was equipped inside the inner cylinder, from which the ER fluid was supplied through a small hole into a clearance between the cylinders. The inner cylinder was fixed to a structure, while the outer cylinder, which had a function as a mass, oscillated along the inner cylinder. When an electric field was applied between the two cylinders, the apparent viscosity of the ER fluid in the clearance varied in response to the applied electric field strength, which caused variation of the damping capacity. Piston rings made of fluoric resin were set up at both ends of the outer cylinder, and were used both as a spacer of two concentric cylinders and as a seal of the ER fluid.

The outer diameter of the inner cylinder was 28 mm, and its length was about 100 mm. The clearance between the two cylinders was designed to be 0.5 mm. The total weight of the present dynamic damper was about 400 g; the weight of the outer cylinder (mass) was 170 g.

The experimental setup is shown in Figure 10. A beam structure was fixed to an electromagnetic exciter at one end, and the present dynamic damper was set up at the free end. The beam was excited in the first and the second natural frequency. Figure 11 shows the responses of the beam at the free end when an electric field was applied statically to the ER dynamic damper. In Figure 11(a), the second natural

vibration appears because the natural frequency of the dynamic damper was set up so that the first mode of the beam might be reduced. As the electric field was strengthened, the first natural vibration mode gradually became conspicuous, as shown in Figures 11(b) and 11(c) (Morishita and Kuroda, 1991).

Vibration Isolator

The last application example is a controllable vibration isolator. A vibration isolator is commonly used for civil engineering structures or precision instruments, such as microscopes.

Figure 12 shows the construction of the ER vibration isolator. The isolator was composed of a top luggage bed, a coil spring, a rubber cover, electrodes, and a diaphragm. The inside of the isolator was filled with ER fluid. The luggage bed was to support a structure on it, and was set up on the coil spring. The coil spring corresponded to the main stiffness of the isolator. The rubber cover was arranged around the coil spring, which had two functions: one was to transmit the volume change (which was caused by the flexibility of the coil spring) of the upper part of the isolator to the lower part through the electrodes, and the other was to prevent the ER fluid from pouring out of the isolator. If the volume change of the upper part of the isolator was not transmitted to the lower part, the ER fluid would not pass through the electrodes, producing no damping. The volume change of the lower part of the isolator was guaranteed by the flexibility of the diaphragm, which was set up at the bottom of the isolator.

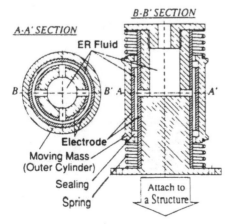

Figure 9. *Damping-controllable dynamic damper.*

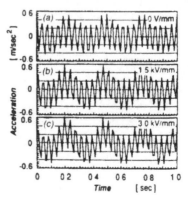

Figure 11. *Response of beam.*

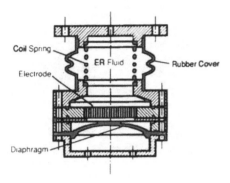

Figure 12. *ER vibration isolator.*

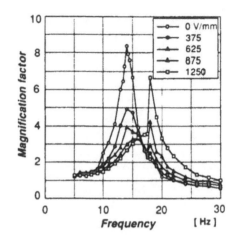

Figure 14. *Response curve (spacing = 0.8 mm, amp. = 0.1 mm).*

The experimental apparatus is shown in Figure 13. A weight was put on the isolator, and the isolator was set up on an electromagnetic exciter. The isolator was excited in a sinusoidal wave by the signal produced by a signal generator. Experiments were conducted under constant relative amplitude between the isolator and the weight to study the amplitude effect on the produced damping.

The experimental results are shown in Figures 14 and 15, at the relative amplitude of 0.1 and 0.5 mm, and at a temperature of 40°C. Without any electric field, the response was almost the same as that of a one-degree-of-freedom system. As the electric field was strengthened, the peak of the response curve became lower and round because the damping was increased. At the same time, the peak shifted to a higher frequency, from 15 Hz to 17 Hz. As the applied electric field was strengthened further, the system damping was expected to become smaller. This might be caused by the membrane force effect of the rubber cover pressurized by the ER fluid inside of the isolator. Comparing the results shown in Figures 14 and 15, the damping capacity was dependent on the vibration amplitude, which was caused by the mechanism of the ER effect. When the vibration amplitude was not so large, the ER effect was strong enough to shut the flow through the electrodes. As the amplitude became large, the chain structures between the electrodes were destroyed by the strong flow through the electrodes (Morishita and Mitsui, 1992b).

ADAPTIVE NEURAL-NET CONTROL SYSTEM

Although various kinds of control theories have been developed, a neural-net control technique is one of the most

suitable for the ER devices, since it is very difficult to develop a conventional control system due to the nonlinearity of the ER effect in regard to electric field strength, temperature or shear rate. The present neural-net control system is based on an adaptive control system, which is composed of a forward model network for structural identification and a controller network (Fujii, Ura and Kuroda, 1990). The architecture is shown in Figure 16. The forward model network represents the dynamics of a structure including the ER devices, and is used for controller optimization. The evaluation and adaptation mechanism has two procedures: one is a forward model modification, and the other is a controller optimization. Adapted by the forward model modification mechanism, the synaptic weights of the forward model can be updated even if the dynamics of the structure or the ER devices change.

As an example, the present control system was applied to the damping-controllable dynamic damper attached to a beam, as shown in Figure 10. The beam was excited in the first and second natural frequencies. Three accelerometers were attached: one at the free end, one at the middle point of the beam, and one on the mass of the dynamic damper.

Figure 17 shows the response spectrum of the experimen-

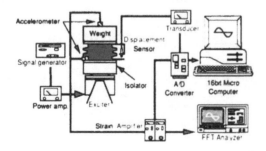

Figure 13. *Experimental setup for vibration isolator.*

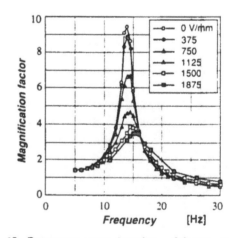

Figure 15. *Resonance curve (spacing = 0.8 mm. amp. = 0.5 mm).*

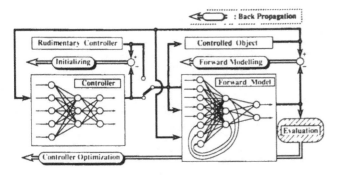

Figure 16. General architecture of SONCS (self-organizing neural-net control system).

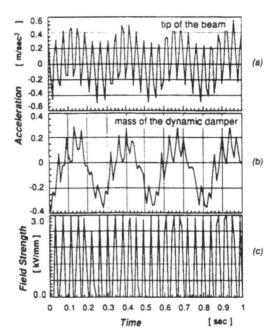

Figure 18. Response and control signal.

tal results. Figure 17(a) shows the acceleration at the end of the beam when the damper was not attached. The result with the initial neural-net controller is shown in Figure 17(b), where only the first natural vibration was reduced (caused by the effect of the conventional dynamic damper). Figure 17(c) shows the response after the adaptation of the present neural-net control system. The first mode as well as the second mode of the beam response were reduced remarkably with a single dynamic damper (Morishita and Kuroda, 1991). The acceleration at the mass of the dynamic damper and the control signal produced by the neural-net controller are shown in Figure 18. As shown in the figure, the acceleration at the mass included both the first and second natural frequencies of the beam by controlling the damping capacity of the dynamic damper.

variation range of viscosity are sufficient for practical use, the fatigue of ER fluid (loss of efficiency), the sedimentation of particles, and the restriction of temperature and shear rate are problems that need to be improved.

CONCLUSIONS

In the present article, four applications of ER fluid are presented: a shock absorber, a squeeze film damper, a dynamic damper and a vibration isolator. All the examples have damping-controllable components, and are applicable for vibration control. The adaptive neural-net control system described here is one of the most suitable control systems for ER actuators.

ER fluid has great potential as a machine element in control systems. Although the quick response time and the wide

REFERENCES

Andrade, E. N. Da C. and C. Dodd. 1946. "The Effect of an Electric Field on the Viscosity of Liquids", *Proceedings of Royal Society A, Vol. 187,* pp. 296–337.

Bjornstahl, Y. and K. O. Snellman. 1937. "Die Einwirkung eines elektrischen Feldes auf die Viskositat bei reinen Flussigkeiten und kolloiden Losungen", *Kolloid Zeitschrift,* 78(3):258–272.

Block, H and J. P. Kelly. 1988. "Electro-Rheology", *Journal of Physics D: Applied Physics,* 21(12):1661 1677.

Bullough, W. A and J. D. Stringer. 1973. "The Utilization of the Electroviscous Effect in a Fluid Power System", *Proceedings of 3rd International Fluid Power Symposium,* pp. 37 52.

Duclos, T. G. 1987. "An Externally Tunable Hydraulic Mount which Uses Fluid", SAE Technical Paper, #870963.

Duff, A. W. 1896. "The Viscosity of Polarized Dielectrics", *Physical Review,* 4(1):23–38.

Fujii, T., T. Ura and Y. Kuroda. 1990. "Development of Self-Organizing Neural-Net-Controller System and Its Application to Underwater Vehicles", *Journal of Society of Naval Architects of Japan,* 168:275–281 (in Japanese)

Inoue, A. 1989. "Study of New Electrorheological Fluid", *Proceedings of Second International Conference on ER Fluid,* pp. 176–183.

Jordan, T. C. and M. T. Shaw. 1989. "Electrorheology", *IEEE Transaction on Electrical Insulation,* 24(5):849–879.

Klass, D. L. and T. W. Martinek. 1967. "Electroviscous Fluids. I. Rheological Properties", "Electroviscous Fluids. II. Electrical Properties", *Journal of Applied Physics,* 38(1):67 80.

Morishita, S. and Y. Kuroda. 1991. "Controllable Dynamic Damper as an Application of Electro-Rheological Fluid", *Proceedings of PVP Conference, ASME, PVP-211,* pp. 1–6.

Morishita, S. and J. Mitsui. 1991. "Controllable Shock Absorber System

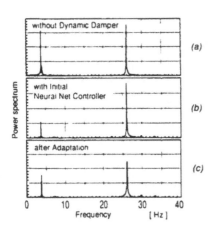

Figure 17. Response spectrum.

(An Application of Electro-Rheological Fluid)", SAE Technical paper #910744.

Morishita, S. and J. Mitsui. 1992a. "Controllable Squeeze Film Damper as an Application of Electro-Rheological Fluid", *Transaction of ASME, Journal of Vibration and Acoustic*, 114(3):354–357.

Morishita, S. and J. Mitsui. 1992b. "An Electronically Controllable Engine Mount Using Electro-Rheological Fluid", SAE Technical Paper #922290.

Petek, N. K., R. J. Goudie and F. P. Boyle. 1988. "Actively Controlled Damping in Electrorheological Fluid-Filled Engine Mounts", SAE paper, #881785.

Petek, N. K. 1992. "An Electronically Controlled Shock Absorber Using Electrorheological Fluid", SAE Technical Paper, #920275.

Stevens, N. G., J. L. Sproston and R. Stanway. 1984. "Experimental Evalu-

ation of Simple Electroviscous Damper", *Journal of Electrostatics*, 15:275–283.

Stevens, N. G., J. L. Sproston and R. Stanway. 1988. "An Experimental Study of Electro-Rheological Torque Transmission", *Transaction of ASME, Journal of Mechanism, Transmission, and Automation in Design*, 110(2):182–188.

Sugimoto, N. and T. Kondo. 1976. "Test-Manufacture of Artificial Fingers", *Research Report of the Research Institute of Industrial Safety*, RIIS-RR-24-8.

Ushijima, T., K. Takano and T. Noguchi. 1988. "Rheological Characteristics of ER Fluids and Their Application to Anti-Vibration Devices with Control Mechanism for Automobiles", SAE Technical paper #881787.

Winslow, W. M. 1949. "Induced Vibration of Suspensions", *Journal of Applied Physics*, 20:1137–1140.

Influence of a Locally Applied Electro-Rheological Fluid Layer on Vibration of a Simple Cantilever Beam

GONG HAIQING,* LIM MONG KING AND TAN BEE CHER

School of Mechanical & Production Engineering
Nanyang Technological University
Nanyang Avenue, Singapore 2263

ABSTRACT: This experimental work is concerned with the vibration characteristics of a cantilever beam locally linked by an electro-rheological fluid layer to ground. Such a novel vibration control strategy can drastically change the vibration characteristics of the beam system, owing to the mechanism that the locally applied ER fluid layer serves as a complex spring and thus change the stiffness matrix, or the structural configuration, of the original beam system under the electric field. It is found that the vibration characteristics of the cantilever beam with the locally applied ER fluid layer treatment is more sensitive to the electric field than a sandwich beam. The influence of the non-linear deformation of the ER fluid due to the presence of the yield stress on the vibration of the cantilever beam is also investigated.

INTRODUCTION

ELECTRO-RHEOLOGICAL fluids (ER) are suspensions of fine particles serving as the dispersed phase in a non-conducting base liquid. As an electric field is applied to the ER fluid, the dispersed phase rearranges itself to form a chain-like structure so that the material properties of the ER fluid change. Such a change in properties is fast, field-dependent, and reversible (Winslow, 1947; Klass and Martinek, 1967; Marshall et al., 1989; Sprecher et al., 1987; Shulman et al., 1989; Block and Kelly, 1988; Conrad et al., 1989). Many mechanical and hydraulic devices in the control technology may soon find an entirely new design alternative involving system characteristics that can be adjusted by using the electric field to meet performance requirements (Stangroom, 1983; Hartsock et al., 1989; Gorodkin et al., 1979; Brooks, 1989; Duclos, 1987, 1988). In particular, research has shown that this material could offer a unique way to tackle many vibration control problems, such as vibration controls of engine mounts (Duclos, 1987; Duclos, 1988; Shulman et al., 1987) and viscoelastic layer treated structures (Coulter and Duclos, 1989; Choi et al., 1990).

Among many investigations related to vibration controls using ER fluids, Coulter et al. (1989) studied the vibration characteristics of a sandwich beam containing an ER fluid. In their study, a transverse excitation was applied to the beam to obtain the resonant frequencies and loss factors of the beam. It was found that the resonant frequencies increase slightly with the electric field strength, and the damping loss factor decreases with mode number and in-

creases with electric field strength. However, the above changes in the vibration characteristics of the sandwich beam are limited. Another experimental investigation on a sandwich cantilever beam was conducted by Choi et al. (1990) in which the vibration decay method is used to investigate the influence of the electric field strength on the vibration damping of the beam. Both the loss modulus and the storage modulus of the sandwich beam increase with the field strength; however, at certain water contents, the loss factor of the beam does not increase with the field strength. Tan (1992) conducted a modal testing of a sandwich beam entirely filled with an ER fluid, and the results showed that the loss factors and resonant frequencies exhibit very small changes under the electric field strength up to 5 (kV/mm).

A rigorous analytical investigation to interpret the above vibration characteristics is difficult to conduct due to the existence of the multi-branch (the pre-yield and post-yield stages) in the Bingham model (Sprecher et al., 1987; Block and Kelly, 1988; Coulter and Duclos, 1989) used to describe the deformations of the ER fluid under an electric field, which causes an overall non-linear deformation of an ER fluid. Based on the Bingham rheological model, the stress response of the ER fluid under a sinusoidal deformation is no longer sinusoidal, as shown experimentally by Yen and Achon (1990), and analytically by Yoshimura and Prud'homme (1987); instead, a flat waveform appears in the response curve in the time domain when the ER fluid yields. On the other hand, the Bingham behavior of the ER fluid in the post-yield stage can be decomposed into a Coulomb damping part and a viscous damping part, and therefore a system damped by the Bingham fluid yields a non-harmonic displacement response under a sinusoidal excitation force (Den Hartog, 1931). However, it is not yet clear how this

*Author to whom correspondence should be addressed.

non-linear deformation affects the overall vibration characteristics of a system.

The non-linear response introduced by the yield stress has posed a great difficulty to the analyses and modeling of the damping mechanism of ERF-based dampers. It is believed that the traditional linear vibration theories, such as the one for the viscoelastic layer treatment, have to be modified to describe the vibration characteristics of an ERF-contained structure. If the excitation applied to an ERF-based system is so small that ER fluid is in its pre-yield stage, then the deformation of the ER fluid is almost linearly elastic. On the other hand, if the excitation strain is much larger than the yield strain of an ER fluid, the non-linearity caused by the yield strain on the vibration of the system becomes limited, as indicated by Den Hartog (1931) in his analysis of a mass-spring system with a mixed damping; thus, the major portion of the damping imposed by the ER fluid is linearly viscous. However, if the excitation strain is in the same magnitude as the yield strain, both pre-yield and post-yield deformations of the ER fluid take equally significant portions within one cycle of the periodic deformation in the time domain; therefore, the system response could be highly non-linear. The above speculations on the effect of the non-linear deformation of the ER fluid become one of the main concerns of this experimental work which shows that the frequency response function curve is indeed affected by the level of excitation force.

The main objective of this work is to find an effective way to change vibration characteristics of a beam using an ER fluid layer. This effort was motivated by our experimental finding (Tan, 1992) that the effect of constrained ERF viscoelastic layer treatment on loss factor and resonant frequency is very limited. A main reason for the weak effect could be that both shear and extensional moduli of the ER fluid under an electric field are not high enough to dissipate a significant amount of energy when the vibration amplitudes of most sandwich ERF beams are very small. However, if a new strategy is adopted so that, rather than serving as a damper, the ER fluid is used to change the structural configuration of the original vibration system, or to change the stiffness matrix of the beam system, a more significant effect in changing the vibration characteristics could be obtained.

In this work, the ER fluid is used mainly as a local spring support, rather than as a damper, to change the stiffness matrix of the cantilever beam system, and therefore the response function as well as the resonant frequencies can be drastically changed. In this approach, the ER fluid layer is applied only to a small portion of a simple cantilever beam and serves as a complex spring used to link the beam to the ground. The electric field strength at a level ranging from 0 to 5 (kV/mm) was applied to the ER fluid layer of 1 (mm) in thickness. The results showed that the frequency response function curve and the resonant frequencies of the beam system change drastically under electric field. The above vibration response was found to be non-linear in the sense that the frequency response function curve changes its shape

at different magnitudes of excitation force. This non-linear vibration characteristic is closely associated with the presence of the yield stress, which imposes a damping of both frictional and viscous nature, as well as an additional stiffness, onto the beam.

EXPERIMENTAL SETUP

As shown in Figure 1, a sinusoidal excitation force generated by a mini-shaker was applied to the mid-section of a cantilever beam, and the output signal was detected by the accelerometer at the free end of the beam. The stainless steel beam was clamped between the top of a supporting column and a clamping plate insulated by two plastic plates. The beam has a Young's modulus of 195.05×10^9 N/m^2 and a density of 7.83×10^9 kg/m^3. The dimensions of the beam are constrained to be 150 mm in length, 15 mm in width, and 0.5 mm in thickness. Such dimensions yield the first natural frequency of 17.9 Hz in the absence of the ER fluid. The ER fluid layer is 1 mm in thickness, 15 mm in width and 10 mm in length, confined in the space between the beam and an aluminium block linked to the ground. Since surface tension is adequate to keep the ER fluid inside the gap, no other material is needed to confine the ER fluid along the gap and the fluid is free to flow between the gap. The ER fluid layer is located at the mid-section of the beam and therefore serves as a complex spring with the stiffness $k^* = k(1 + i\eta)$. In this arrangement, the ER fluid layer is under compression/extensional deformation. Four different electric field strengths were introduced to change the vibration property of the ER fluid and hence the beam system. Three different levels of excitation forces, or power gains, were introduced to investigate the non-linear properties of the ERF-based system.

The mini-shaker has the specified frequency response from 10 Hz to 20 kHz. The frequency sweep of the sinusoidal excitation is taken directly from the FFT analyzer, which has a built-in signal generator that can output a random signal when measuring the transfer function of the beam system. Three positions of the power gains in the power amplifier were selected for the experiment within the linear range of the vibration characteristics of the cantilever

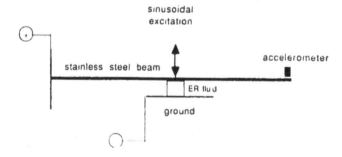

Figure 1. Test rig configuration of a cantilever beam with an ER fluid layer applied at the mid-section of the beam.

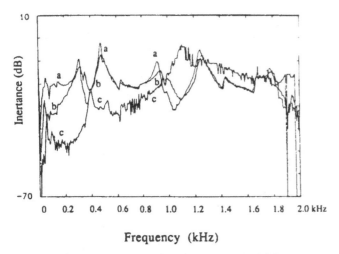

Figure 2. *Frequency response function curve of the ERF cantilever beam at excitation level (gain) I. Curve a:* E = 0 (kV/mm); *curve b:* E = 1 (kV/mm); *curve c:* E = 3 (kV/mm).

beam when the ER fluid is absent. The DC high voltage generator can supply a continuously variable DC voltage from 0 to 5 kV. The ER fluid prepared by Lord Corporation has the dynamic yield stress of 859 Pa at the field strength of 3 kV/mm. This experiment was conducted at the constant temperature of 24°C.

RESULTS AND DISCUSSIONS

Effect of the Electric Field

Various electric field strengths from 0 to 5 kV/mm are chosen to examine their influence on the vibration characteristics of the cantilever beam. Due to the non-linearity introduced by the yield behavior of the ER fluid, the above influence of the electric field on the beam vibration varies with the magnitude of the excitation force. In this work, three power gains are introduced to reveal the effect of the non-linear deformation of the ER fluid. Gain I is at the gain position 1 in the B&K power amplifier Type 2712, gain II is at the gain position 3, and gain III is at the gain position 5. The magnitudes of these gains are chosen so that the vibration of the beam is linear when the ER fluid layer is absent. In this study, the inertance (acceleration per unit force) is used as the response function.

Figure 2 shows the influence of the electric field strength on the frequency response function curve of the cantilever beam at gain I. The frequency response function curve changes significantly with the field strength [compared to those of the sandwich beam obtained by Tan (1992)]. When the field strength increases from $E = 0$ kV mm to $E = 1$ kV mm, the frequency response function curve does not change as much as that compared to the response at an electric field strength of $E = 3$ kV/mm. When the field strength increases to $E = 1$ kV/mm, the resonant frequencies increase slightly. As the field strength increases to $E =$

3 kV/mm, some of the resonances are greatly attenuated and the shape of the frequency response function curve is significantly altered, indicating a change in the stiffness matrix of the beam system under the electric field.

It is further observed in this work that the frequency response function curve shifts downward to yield lower resonant peak levels and higher resonant frequencies as the field strength increases. Such changes in the frequency response function curve are due to the increasing stiffness of the system, rather than to the change of the damping loss factor (Nashif et al., 1985). Such a downward-shift is in accordance with the similar downward-shift of the compliance curve due to an increasing stiffness in a typical single-degree freedom system consisting of a mass and a complex spring under a forced excitation (Nashif et al., 1985). At $E = 0$ kV/mm, the ERF-based beam system may experience a linear behavior since the ER fluid is still in its liquid form. At $E = 1$ kV/mm, the ER fluid may still behave linearly, i.e., the ER fluid behaves more or less as a simple liquid damper, and the effect of the yield stress is significant only at the first resonance. At the high field strength, $E = 3$ kV/mm, the gain level of the excitation force applied may not be sufficient to cause the ER fluid to yield in the major portion of a cycle under the sinusoidal excitation. Therefore, the increasing stiffness of the ER fluid causes a change of each element in the stiffness matrix of the beam system, which results in a drastic change in shape of the frequency response function curve. Figure 3 shows the resonant peak levels of the first three modes (also see Figure 2) at different field strengths at the power gain I. It is shown that the resonant peak levels decrease significantly with the increasing field strength for the first three modes. The increase of the resonant frequencies for the first three modes with the field strength is shown in Figure 4, which indicates the increasing stiffness of the system as the field strength increases.

The influence of the field strength at gain I on the loss factors for the first three modes is shown in Figure 5. The changes of the loss factors as the function of the field

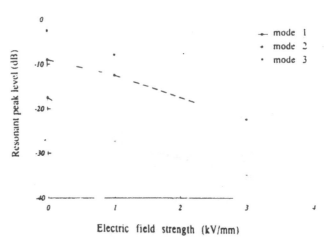

Figure 3. *Resonant peak levels vs. electric field strength at excitation level (gain) I.*

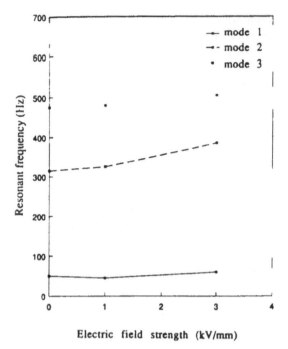

Figure 4. *Resonant frequencies vs. electric field strength at excitation level (gain) I.*

strength are not monotonic, and such changes are small compared to the changes of the resonant peak levels and the resonant frequencies, which indicates that the loss factor takes a less important role in designing a control strategy.

Figures 6 and 7 show the frequency response function curves at the power gains II and III with the electric field strengths $E = 0$ kV/mm, $E = 3$ kV/mm, and $E = 5$

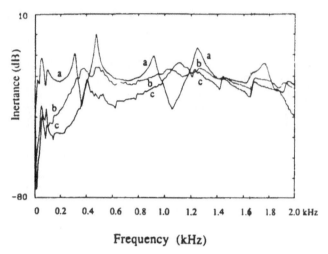

Figure 6. *Frequency response function curve of the ERF cantilever beam at excitation level (gain) II. Curve a:* E = 0 *(kV/mm); curve b:* E = 3 *(kV/mm); curve c:* E = 5 *(kV/mm).*

kV/mm. It is shown that frequency response function curves at gain II and III bear similar features to the one at gain I.

Effect of Non-Linear Deformation of the ER Fluid

The rheology that an ER fluid exhibits can approximately be described by the Bingham rheological model in which a field-dependent yield stress is incorporated. Yoshimura and Prud'homme (1987) analytically showed that a Bingham material does not yield a sinusoidal stress response under a sinusoidal deformation, and the same behavior was experimentally observed by Yen and Achon (1990). Under a sinusoidal deformation, the stress response of the ER fluid exhibits a flat waveform in its response curve in the time domain. The flat portion of the response curve appears when the ER fluid yields. Therefore, a non-linearity characterised by the yield stress is introduced to the system containing the ER fluid. The effect of such a non-linearity on a vibration

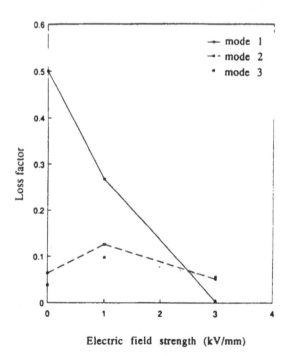

Figure 5. *Damping loss factors vs. electric field strength at excitation level (gain) I.*

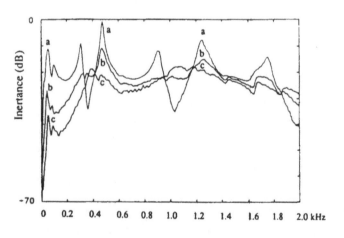

Figure 7. *Frequency response function curve of the ERF cantilever beam at excitation level (gain) III. Curve a:* E = 0 *(kV/mm); curve b:* E = 3 *(kV/mm); curve c:* E = 5 *(kV/mm).*

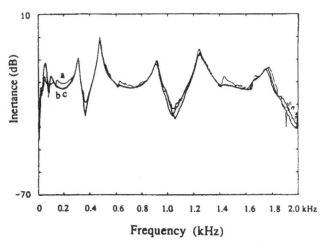

Figure 8. *Frequency response function curve of the ERF cantilever beam at E = 0 (kV/mm). Curve a: gain I; curve b: gain II; curve c: gain III.*

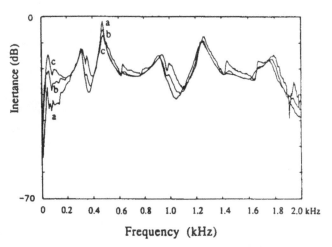

Figure 9. *Frequency response function curve of the ERF cantilever beam at E = 1 (kV/mm). Curve a: gain I; curve b: gain II; curve c: gain III.*

system is complex and depends on the amplitude of the displacement of the beam vibration relative to the yield strain of the ER fluid. In other words, the non-linear effect depends on the portion taken by the pre-yield deformation and post-yield deformation within one cycle of the sinusoidal deformation of the ER fluid. As indicated by Den Hartog (1931) in his analysis on a mass-spring system with a mixed friction-viscous damping, if an excitation amplitude is much larger than the yield strain of a fluid, the effect of the yield strain on the frequency response function is limited. When the excitation amplitude is large, the major portion of the sinusoidal deformation of the ER fluid is in its post-yield liquid state; therefore, the damping imposed by the ER fluid is approximately linearly viscous.

At $E = 0$ kV/mm, the yield strain is nearly zero; therefore, the ER fluid is merely liquid damper in the beam system. It can be observed that the frequency response function curve has only slight changes at low frequencies at different gains of the excitation force, as shown in Figure 8, in which the vibration characteristics of the beam system appear to be linear. As the electric field strength increases to $E = 1$ kV/mm, the yield strain becomes non-zero and the non-linear effect is observed at lower frequencies, as shown in Figure 9. At this field strength, the frequency response function curve is almost the same as that in Figure 8, with some small differences at low frequencies. This characteristic is due to the dominating post-yield deformation at $E = 1$ kV/mm because of a relatively low yield strain; therefore, the ER fluid still behaves like a liquid damper except at the low frequencies.

As the field strength further increases to $E = 3$ kV/mm, the frequency response function curve changes entirely at different gains, as shown in Figure 10. At this field strength, both the pre-yield and post-yield deformations play significant roles in affecting the system vibration, and a high non-linearity is introduced to the beam system by the ER fluid; it can no longer behave like the liquid damper. It is also evi-

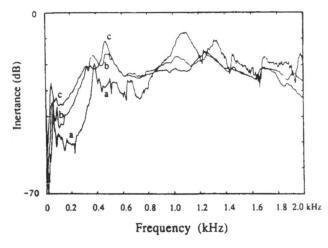

Figure 10. *Frequency response function curve of the ERF cantilever beam at E = 3 (kV/mm). Curve a: gain I; curve b: gain II; curve c: gain III.*

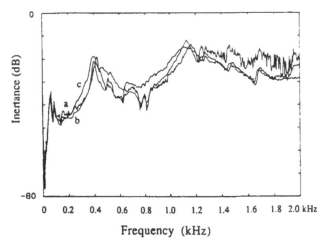

Figure 11. *Frequency response function curve of the ERF cantilever beam at E = 5 (kV/mm). Curve a: gain I; curve b: gain II; curve c: gain III.*

dent that the variation of the frequency response function curves with different power gains at $E = 5$ kV/mm is more like that at $E = 0$ kV/mm or $E = 1$ kV/mm than that at $E = 3$ kV/mm. At $E = 5$ kV/mm, the frequency response function curves at different gains converge and the non-linear effect becomes much weaker, as shown in Figure 11. This finding indicates that the ER fluid behaves like a linear spring in the beam system, and the pre-yield deformation becomes dominant in the ER fluid due to the higher yield strain. As a result, the beam system partially recovers its linearity.

CONCLUSIONS

A new strategy of controlling structural vibrations using ER fluids was experimentally investigated. The ER fluid layer was locally applied to a cantilever beam as a complex spring. Such a short strip of ER fluid layer changed the structural configuration of the original beam system or the arrangement of the stiffness matrix of the beam system when an electric field was applied. It was found in this experiment that the frequency response function curve of the ER beam system changed drastically under an electric field. These changes include a downward shift of the frequency response function curve, significant increase in resonant frequencies, change of damping loss factors, etc. The above changes, as compared with experimental observations on a sandwich beam entirely filled with the ER fluid, are due to the change in the stiffness matrix of the beam system under the electric field, rather than to a change in damping loss factor. In addition, the above vibration response was found to be non-linear in the sense that the frequency response function curve changes its shape at different gains of excitation force. This non-linear vibration characteristic is closely associated with the role of the yield stress, which also imposes a damping of both frictional and viscous nature alternately within one cycle of the sinusoidal deformation of the ER fluid, as well as an additional stiffness, onto the beam.

REFERENCES

Block, H. and J. P. Kelly. 1988. "Electro-Rheology", *J. Phys. D: Appl. Phys.*, 21:1661–1677.

Brooks, D. A. 1989. "Devices Using Electro-Rheological Fluids", *Proc. 2nd Int. Conf. on ER Fluids, Raleigh, NC*, pp. 371–401

Choi, Y., A. F. Sprecher and H. Conrad. 1990. "Vibration Characteristics of a Composite Beam Containing an Electrorheological Fluid", *J. Intelligent Material Systems and Structures*, 1:91–104.

Conrad, H., Y. Chen and A. F. Sprecher. 1989. "Electrorheology of Suspensions of Zeolite Particle in Silicone Oil", *Proc. 2nd Int. Conf. on ER Fluids*, pp. 252–264.

Coulter, J. P. and T. G. Duclos. 1989. "Applications of Electrorheological Materials in Vibration Control", *Proc. 2nd Int. Conf. on ER Fluids, Raleigh, NC*, pp. 300–325.

Den Hartog, J. P. 1931. "Forced Vibration with Combined Coulomb and Viscous Friction", *Trans. A.S.M.E.*, APM-53-9, pp. 107–115.

Duclos, T. G. 1987. "An Externally Tunable Hydraulic Mount which Uses Electro-Rheological Fluid", *Proc. Noise and Vibration Conf., Traverse City, MI*, pp. 131–137.

Duclos, T. G. 1988. "Design of Devices Using Electrorheological Fluids", *Proc. Future Transportation Technology Conf., San Francisco, CA*, pp. 1–5.

Gorodkin, R. G. et al. 1979. "Applications of the Electrorheological Effect in Engineering Practice", *Fluid Mech.-Soviet Research*, 8(4):48–61.

Hartsock, D. L., R. F. Novak and G. J. Chaundy. 1989. "ER Fluid Requirements for Automotive Devices", *J. Rheology.*

Klass, D. L. and T. W. Martinek. 1967. "Electroviscous Fluids. I. Rheological Properties", *J. of Applied Physics*, 38(1):67–74.

Marshall, L., C. F. Zukoski and J. W. Goodwin. 1989. "Effects of Electric Fields on the Rheology of Non-Aqueous Concentrated Suspensions", *J. Chem. Soc., Faraday Trans.*, 85(9):2785–2795.

Nashif, A. D., D. I. G. Jones and J. P. Henderson. 1985. *Vibration Damping*. New York, NY: John Wiley & Sons.

Shulman, Z. P. et al. 1987. "Damping of Mechanical-Systems Oscillations by a Non-Newtonian Fluid with Electric-Field Dependent Parameters", *J. Non-Newtonian Fluid Mech.*, 25:329–346.

Shulman, Z. P., E. V. Korobko and Yu. G. Yanovskii. 1989. "The Mechanism of the Viscoelastic Behaviour of Electrorheological Suspensions", *J. of Non-Newtonian Fluid Mech.*, 33:181–196.

Sprecher, A. F., J. D. Carlson and H. Conrad. 1987. "Electrorheology at Small Strains and Strain Rates of Suspensions of Silica Particles in Silicone Oil", *Mat. Sci. Engr.*, 95:187–197.

Stangroom, J. E. 1983. "Electrorheological Fluids", *Phys. Technol.*, 14:290–296.

Tan, B. C. 1992. "An Experimental Investigation of Electro-Rheological Fluid Based Vibration Control System", *FYP Report*. Singapore: Nanyang Technological University.

Yen, W. S. and P. J. Achon. 1990. "A Study of the Dynamic Behavior of an Electro-Rheological Fluid", presented at the 62nd Annual Meeting of the Society of Rheology. Santa Fe, NM.

Yoshimura, A. S. and R. K. Prud'homme. 1987. "Response of an Elastic Bingham Fluid to Oscillatory Shear", *Rheol. Acta*. 26:428–436.

Winslow, W. M. 1947. U.S. patent 2,427,850, March 25.

Electrorheological Material under Oscillatory Shear

J. H. Spurk*
Technische Hochschule Darmstadt
Petersenstr. 30
6100 Darmstadt
Germany

Zhen Huang
August Bilstein GmbH
Postfach 1151
5828 Ennepetal
Germany

ABSTRACT: Two typical electrorheological materials, which are anhydrous dispersions of silica particles in silicone oil with surface active additives, are sheared in oscillatory motion in a gap, formed by an outer fixed cylinder and an inner, axially movable cylinder. The rheometer is operated at tuneable resonance frequencies, such that only the reaction of the material is measured. Long time oscillatory shear measurements show that the electroviscous stress deteriorates under d.c. electric field, while electroviscous stress is stable when a.c. fields are used. Electroviscous stress is found to depend on field frequency through the dependence of the dielectric constant of the disperse phase on field frequency. A dimensionless quantity, which relates the electroviscous shear stress to the electric field strength, serves to correlate all the experimental data for different shear amplitudes and shear frequencies for each fluid. The phenomenological theory of Huang and Spurk predicts this efficiency with satisfactory accuracy.

INTRODUCTION

SINCE W. M. Winslow's discovery (Winslow, 1949) that the apparent viscosity of a suspension of dielectric liquid and solid particles can be changed by an external electric field, interest in this "electroviscous" or "electrorheological" material has been kept alive because of the many applications such an easily controlled material could have. The materials exhibit Bingham-behaviour under electric field, and the field strength dependent yield stress gives rise to the apparent viscosity, or to its difference from the zero field viscosity, the electroviscosity. The materials, available in the past, however, have not shown a sufficiently high yield stress to find significant commercial use. In this article, we report experiments on two experimental materials thought to be typical of the new generation of liquid electrorheological materials of low conductivity and with extended temperature range, whose properties are largely determined by surface active additives and which show an enhanced electroviscous effect. We also introduce an objective measure to characterize electrorheological material, a need that has been apparent for some time (Anonymous, 1988).

The ease with which the viscosity or shear stress can be influenced is especially attractive in applications to vibration control in automotive and aerospace applications. In most of the devices considered, the fluid is subjected to oscillatory shear. It is therefore desirable to know the material behaviour under oscillatory shear; we place emphasis on these results, but also report on the electroviscosity under steady shear. The experimental determination of the aforementioned measure, an efficiency which relates the electroviscous stress to the expended field strength, is the most important result. For the conditions of our experiments, this efficiency can be obtained from the phenomenological theory of Huang and Spurk (1990) up to an unknown length ratio, which is adjusted to fit the experiments.

MATERIALS USED

The two electrorheological fluids used in this work and later referred to as ERF1 and ERF2 are dispersions of silicone oil, silica particle and surface active additives in the volume fractions 65%/31.9%/3.1% and 63%/33%/3.3%, respectively. The primary particle size is about 0.1 μm. The electrorheological fluids are not water based, but rather the additives are chemically bound (surface grafted) to the surface. The particles aggregate and form secondary macrostructures of about 10 μm. Under steady shear, the macrostructures can be damaged or destroyed and the suspension is shear thinning. Even though the composition of the two fluids is nearly the same, they display markedly different material behaviour, indicative of the active role the additives have on the suspension. In Figure 1, the effective, specific

*Author to whom correspondence should be addressed

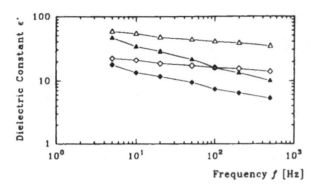

Figure 1. Dielectric constant ϵ' of the dispersion at $\dot{\gamma}$ = 200 1/s and E = 1 kV/mm as function of field frequency f (data supplied by manufacturer). Dielectric constant ϵ_k' of the disperse phase computed by Equation (1)—ERF1: ▲ ϵ_k', ◆ ϵ'; ERF2: ϵ_k', ◇ ϵ'.

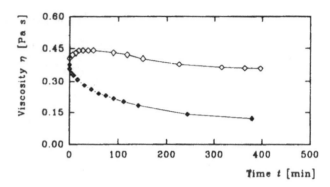

Figure 2. Zero field viscosity at constant shear ($\dot{\gamma}$ = 200; 1/s), filled symbols: ERF1, open symbols: ERF2.

dielectric constants $\epsilon' = \epsilon/\epsilon_0$, where ϵ_0 is the vacuum dielectric constant, are plotted as functions of electric field frequency. Note that the ERF1 has a substantially lower dielectric constant than the ERF2 (the main reason for it being electrorheologically less efficient), and that if falls off more rapidly with field frequency. For later use, the figure also shows the effective (i.e., including the surfactant's effect) dielectric constant ϵ_k' for the particles computed with the dielectric constant of the carrier fluid $\epsilon_f' = 2.7$ from the linear relation

$$\epsilon' = \Phi\epsilon_k' + (1 - \Phi)\epsilon_f' \qquad (1)$$

in which Φ is the volume fraction of the solid material. We use this explicit relation since the Bruggeman formula used by Klass and Martinek (1967) for imbedded spherical particles cannot be solved for ϵ_k' for all conditions of interest. Equation (1) is the Wiener limit, which gives the upper bound for ϵ' for all possible aggregates of granular and rod-like particles in the composition $\Phi/(1 - \Phi)$ (Bruggeman, 1935). The dielectric constants can also be modified by shear. While the ERF2 shows almost no shear dependence at a field frequency of 10 Hz and 1 kV/mm field strength under steady shear, the ERF1 suffers a decay of the dielectric constant of nearly 100% up to shear rates of 500 s⁻¹. Most likely, the decay is caused by the destruction of the macrostructure, since the flow modified permittivity (FMP) due to particle spin, as discussed by Block and Kelly (1989), would not seem to be important with macrostructures intact and at high volume fractions. Long time measurements of the zero field viscosity confirm the difference in the two materials, as shown in Figure 2. The viscosity of the ERF1 decreases with time, while the viscosity of the ERF2 actually increases at first, apparently because the shear favours the formation of secondary structures. The slight subsequent decrease could be caused by sedimentation or changes in the macrostructure. No influence of a change in the macrostructure is visible in the data for oscillatory shear given in Figure 3, where there is no degradation with time. The complex dependence on amplitude A_m and shear frequency f, suggests that sometimes the primary particles just slide within the macrostructure, while at other times, especially at larger amplitudes, the secondary structures move as

a whole. Generally, without field the viscosity increases with increasing amplitude but the dependence on frequency is not monotonous (Figure 4). The material behaviour under an external field is more transparent—and more easily controlled—than the zero field behaviour, because of the order brought about by the external field. We conjecture that under external field, the primary particles orient themselves in the secondary macrostructures, which then fibrillate and form the familiar columns in the field direction (Figure 5). The size and stability of the secondary structure is controlled by the additives, which not only increase the dielectric constants, but also increase the effective particle size. Both actions contribute to an increase in the electrorheological effect.

APPARATUS

Since the frequencies in the applications mentioned lead to high inertial forces in conventional rheometers, a rheometer has been built whose inertial forces are compensated by spring forces. It consists (Figure 6) of a fixed and electrically grounded outer cylinder and an inner movable cylinder to which the high voltage is applied. The outer and inner cylinder form an annular gap, whose height D is much smaller than the radius. The height can be varied between 0.15 to 1.5 mm by using different inner cylinders. The annular gap is sealed to the outside by a thin membrane of

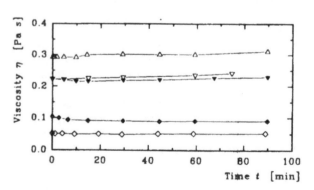

Figure 3. Long time behaviour of zero field viscosity, oscillatory shear; filled symbols: ERF1, open symbols: ERF2. Test conditions: shear frequency f, Hz; amplitude A_m mm, ▼ 51˙ 0.656, ◆ 37; 0.280, 48; 0.656, 142; 0.656, ◇ 31; 0.656.

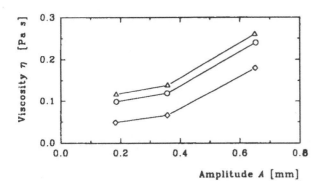

Figure 4. *Dependence of zero field viscosity on amplitude and shear frequency. ERF2. Test condition: shear frequency f_s Hz; ◇ 31;   △ 48;   △ 154.*

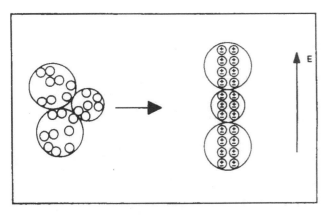

Figure 5. *Sketch of primary and secondary particle structures.*

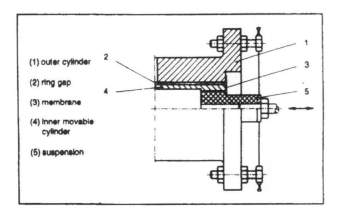

(1) outer cylinder

(2) ring gap

(3) membrane

(4) inner movable cylinder

(5) suspension

Figure 6. *Sketch of rheometer.*

negligible stiffness. The inner cylinder is suspended by piano wires, whose tension and thickness can be adjusted, and it is only capable of moving in the axial direction. The inner cylinder and the wires constitute a mass-spring system whose natural frequency can be tuned to the operating shear frequency f_s by adjusting the piano wires. The system is excited to a steady state oscillatory motion by an electrodynamic shaker having a sinusoidal voltage input. The force necessary to sustain the motion is measured with a piezoelectric force transducer. At resonance, the force necessary for an oscillatory steady state motion, without liquid in the shear gap, is typically 1%´ of the inertia forces. From the force difference at constant frequency and constant amplitude of the motion, with and without liquid, we compute the shear stress under zero field conditions, and from it the viscosity. This computation requires knowledge of the shear rate at the moving wall, say $\dot{\gamma} = (\partial u/\partial y)_w$. We compute the shear rate at the wall on the basis of planar oscillatory Newtonian shear flow (e.g., Spurk, 1989). This is not the true shear rate, which in general is unknown, but a reference quantity. (It would be the true shear rate if the material was Newtonian.) For most cases, the diffusion time $\varrho D^2/\eta$ is much smaller than the characteristic time of the motion $(2\pi f_s)^{-1}$, so that $(2\pi f_s \varrho/\eta)D^2 \ll 1$, and then the shear rate $\dot{\gamma}$ is simply U/D, where U is the velocity of the wall. From the difference of force amplitude with and without electric field, we compute the electroviscous shear stress Σ_{elc} and, by dividing the electroviscous shear stress amplitude by the amplitude of the reference shear rate at the wall, we obtain the electroviscosity η_{elc}, which therefore is a fictitious quantity, since the actual shear rate at the wall is unknown. Since the material is nonlinear, the motion and the force are not sinusoidal and it is not appropriate to introduce a complex viscosity. We therefore find it more consistent to work with the electroviscous stress. The electroviscosity η_{elc} and indeed the viscosity η as used in this article, are the stress amplitude divided by the amplitude of the reference shear rate. A "phase" of nearly 90° and almost constant, between the motion and the force, was observed, but the resolution of the oscillograph traces was too small to reliably measure a systematic trend on amplitude and frequency.

MEASUREMENTS WITH ELECTRIC FIELD

Long time measurements of the electroviscosity with direct current (d.c.) field show the electroviscosity to decrease with time (Figure 7). The rate of decrease is found to depend on the gap height D and on the field strength E, and we conclude that the decrease is due to migration of charged particles to, and accumulation at, the walls in a sedimentation-like process. There is no deterioration of electroviscosity under alternating current (a.c.) fields (Figure 8). For this reason, further measurements were restricted to a.c. fields and carried out for the shear frequencies $f_s \approx 30$, 50, 150 Hz. The electroviscosity depends strongly on the frequency f of the electric field and for fixed shear amplitude A_m, shear frequency f_s, and gap height D, the data can be fitted to a one parameter family of curves

$$\eta_{elc} = \alpha(f, f_s, A_m, D)\epsilon_0 E^2 \qquad (2)$$

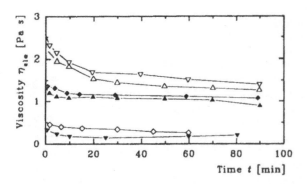

Figure 7. *Illustrating the time dependence of electroviscosity for oscillatory shear and d.c. electric field* ($f_s = 37 + 150$ Hz, $A_m = 0.28 + 0.66$ mm, $D = 0.5 + 1.5$ mm, $E = 1.7 + 4$ kV/mm); *filled symbols: ERF1, open symbols: ERF2.*

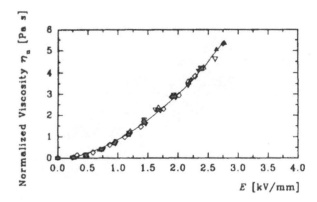

Figure 9. *Normalized viscosity (ERF2, $f_s = 48.3$ Hz, $A_m = 0.656$ mm, $f \sim 1 + 500$ Hz).*

where $\alpha(f)$ is the parameter, and the dependence on f_s, A_m and D is implied. A "normalized" electroviscosity can be introduced when the value of α at a reference frequency of, say, $f = 100$ Hz, is used as a reference value:

$$\eta_n = \frac{\alpha(100)}{\alpha(f)} \eta_{ele} \qquad (3)$$

The normalized electroviscosity is independent of field frequency, as Figure 9 shows. This suggests that Equation (2) has the special form

$$\eta_{ele} = K(f_s, A_m, D)\lambda(f)\epsilon_0 E^2 \qquad (4)$$

where λ is dimensionless and $K(f_s, A_m, D)$ has the dimension of time, which restricts the dependence on the variables for dimensional reasons to $K = f_s^{-1} fn(D/A_m)$. It is natural to try the relation

$$K = f_s^{-1}\frac{D}{A_m} \qquad (5)$$

whose R.H.S., up to an irrelevant constant factor which can

be absorbed in $\lambda(f)$, is equal to $1/\dot{\gamma}$, so that instead of Equation (4) we obtain

$$\Sigma_{ele} = \eta_{ele}\dot{\gamma} = \lambda(f)\epsilon_0 E^2 \qquad (6)$$

or, in dimensionless form,

$$\lambda(f) = \frac{\Sigma_{ele}}{\epsilon_0 E^2} \qquad (7)$$

which is an "efficiency" giving the electroviscous shear stress in terms of the electric field strength necessary to produce it. Of course, this is not an efficiency in the common usage of the word, since it can be larger than one. The function λ, being dimensionless, can only be a function of a dimensionless frequency and we choose, in Figures 10 and 11, the frequency at which λ takes its maximum value as the reference frequency. We will show later that the dependence of λ on field frequency can be traced to the dependence of ϵ' and ϵ_k' on field frequency. The specific dielectric constants must themselves be functions of dimensionless variables, so that the representation in Figure 1 is dimensionally inconsistent. If the dependence of ϵ_k' on f results from a polarisation process, characterized by a relaxation time,

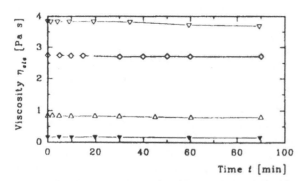

Figure 8. *Illustrating the time dependence of electroviscosity for oscillatory shear and a.c. electric field* ($f_s = 25 + 154$ Hz, $A_m = 0.280 + 0.656$ mm, $D = 0.5$ mm, $E = 1.2$ kV/mm, $f = 100$ Hz); *filled symbols: ERF1, open symbols: ERF2.*

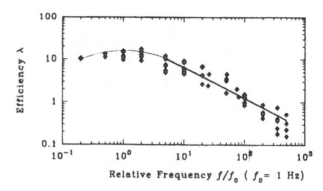

Figure 10. *Efficiency λ as function of relative field frequency for ERF1. — Theory, ♦ Experiments* ($f = 25 + 170$ Hz, $A_m = 0.28 + 0.66$ mm, $D = 0.5$ mm)

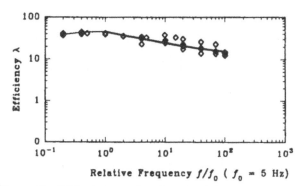

Figure 11. *Efficiency λ as function of relative field frequency for ERF2, — Theory, ◇ Experiments (f_s = 25 + 177 Hz, A_m = 0.28 + 0.66 mm, D = 0.15 + 0.5 mm).*

then it would serve as the natural time to render the frequency in Figure 1 dimensionless.

DISCUSSION AND COMPARISON WITH THEORY

Returning to Figures 10 and 11, which represent the condensed form of our experiments, we observe that the ERF2 with a maximal value $\lambda_{max} \approx 45$ is electrorheologically more efficient than the ERF1 with $\lambda_{max} \approx 17$.** The liquids tested by Klass and Martinek (1967) have efficiencies of $\lambda \approx 2.8$, $\lambda \approx 11.0$ and $\lambda \approx 81.5$ at volume fractions of $\Phi = 0.2$, $\Phi = 0.36$ and $\Phi = 0.46$. Since the electroviscous stress is, according to Equation (6), independent of shear frequency f, and amplitude A_m, the oscillatory shear flow is quasisteady, i.e., at each instant, the flow corresponds to a steady shear, to which the model for the electroviscous effect of Huang and Spurk (1990) is applicable. According to this model, the electroviscous stress results from the work that is necessary to change the particle configuration. The phenomenological theory assumes that the fibers are columns of rectangular "brick"-like particles. The electroviscous shear stress is given by

$$\Sigma_{ele} = \frac{k}{4} \epsilon E^2 \Phi \frac{D}{a} \left(\frac{1}{G_f} + \frac{1}{G_k} - \frac{1}{G_k + G} - \frac{1}{G - G} \right)$$

(8)

where, in obvious notation,

$$G_{f,k} = 1 - H\left(1 - \frac{\epsilon}{\epsilon_{f,k}}\right) {}'D$$

(9)

and

$$G = 2h_k\left(\frac{\epsilon}{\epsilon_f} - \frac{\epsilon}{\epsilon_k}\right) {}'D$$

(10)

In Equation (8), a is the primary particle size, and the

meaning of the geometrical quantities can be taken from Figure 12. The column height H depends implicitly on Σ_{ele} and is determined from the condition that the shear stress from Equation (8) is transmitted as viscous shear stress Σ_s through shear zones near the plates. However, for sufficiently high electric fields and for $h_k/D \ll 1$, H and D vanish from the result:

$$\Sigma_{ele} = \frac{k}{2} \epsilon_f E^2 \Phi \left(1 - \frac{\epsilon_f}{\epsilon_k}\right)\left(\frac{\epsilon_k^2}{\epsilon_f^2} - 1\right)\frac{h_k}{a}$$

(11)

It is now easy to obtain the expression

$$\lambda(f) = \frac{\Sigma_{ele}}{\epsilon_0 E^2} = \frac{k}{2} \epsilon_f'\Phi\left(1 - \frac{\epsilon_f'}{\epsilon_k'}\right)\left(\frac{\epsilon_k'^2}{\epsilon_f'^2} - 1\right)\frac{h_k}{a}$$

(12)

which shows that the dependence on field frequency enters through the dependence of ϵ_k' on the frequency. In Equation (12), k is a dimensionless constant introduced by Huang and Spurk (1990) as a particle shape factor. It was there adjusted to $k = 0.28$ to fit the experiments of Klass and Martinek (1967). The only other geometric quantity in Equation (12) is the ratio h_k/a. It was set equal to one by Huang and Spurk (1990), since the fluids of Klass and Martinek had no recognizable macrostructure. We favor an interpretation of h_k/a as the ratio of the primary to secondary particle size but deduce from our experiments only the product kh_k/a, which was found to be $kh_k/a = 0.08$ for ERF1 and for ERF2, $kh_k/a = 0.21$. The theoretical functions $\lambda(f/f_0)$ determined with these choices are plotted as thick lines in Figures 10 and 11 in the range for which the dielectric constants for the particles are available from Figure 1. The thin lines are extrapolations. The generally good agreement between Equation (12) and the experiments justifies the use of Equation (12) or Equation (8) to show explicitly how the size of the secondary structure enters the efficiency, and the decisive role that is played by the effective dielectric constant of the dispersed phase. Of course, this equation should only be used as a guide, since for its derivation the fiber configuration was postulated rather than determined, as it should be in a consistent theory. In addition, the model of Huang and

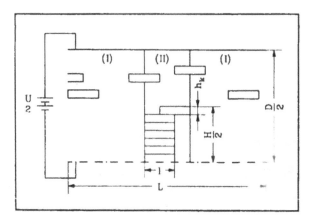

Figure 12. *Sketch of particle column and shear zone above symmetry line.*

Spurk (1990) is restricted to the strong field assumption, i.e.,

$$\frac{\epsilon E^2}{\eta \dot{\gamma}} \gg 1 \tag{13}$$

If $\lambda \gg 1$ (see Figure 10 and Figure 11), we have

$$\lambda \frac{\epsilon E^2}{\eta \dot{\gamma}} \gg \lambda \gg 1 \tag{14}$$

and therefore

$$\frac{\epsilon \eta_{clc}}{\epsilon_0 \eta} \gg 1 \tag{15}$$

and because $\epsilon/\epsilon_0 > 1$ always, also

$$\frac{\eta_{clc}}{\eta} \gg 1 \tag{16}$$

as the sufficient condition for the strong field assumption. This condition is met in our experiments, except for ERF1 at high field frequencies, where the deviation from theory is noticeable.

ACKNOWLEDGEMENTS

This work was supported by Metzeler Gimetall AG, who also supplied the electrorheological fluids. We wish to thank Drs. V. Härtel and W. Michel for their advice and generous help during the course of this study.

REFERENCES

Anonymous. 1988. "Electrorheological Fluids and Devices", *Automotive Engineering*, 96(12):45–48.

Block, H. and J. P. Kelly. 1989. "Materials and Mechanisms in Electro-Rheology", in *Proceedings of the First International Symposium on Electrorheological Fluids*, H. Conrad, A. F. Sprecher and J. D. Carlson, eds., Raleigh, NC: North Carolina State University, Engineering Publications.

Bruggeman, D. A. G. 1935. "Berechnung verschiedener physikalischer Konstanten von heterogenen Substanzen", *Annalen der Physik, Band 24*.

Huang, Z. and J. H. Spurk. 1990. "Der elektroviskose Effekt als Folge elektrostatischer Kraft" *Rheol. Acta*, 29:475–481.

Klass, D. L. and T. W. Martinek. 1967. "Electroviscous Fluids," *J. Applied Physics*, 38(1):67–80.

Spurk, J. H. 1989. *Strömungslehre, 2. Auflage*. Berlin, Heidelberg, New York, London, Tokyo, Hong Kong: Springer Verlag.

Winslow, W. M. 1949. "Induced Vibrations of Suspensions," *J. Applied Physics*, 20:1137–1140.

Fluid Durability in a High Speed Electro-Rheological Clutch

A. R. Johnson,* J. Makin, W. A. Bullough, R. Firoozian and A. Hosseini-Sianaki

Department of Mechanical and Process Engineering
University of Sheffield
Mappin Street
Sheffield, S1 3JD
United Kingdom

ABSTRACT: The durability of an electro-rheological (ER) fluid was investigated by running a high speed ER clutch under different conditions and periods of operation. The tests involved running the clutch at 3000 rpm for a total period of twelve hours over a five day period. The tests subjected the fluid to a centripetal acceleration of 3000 m/s^2, and were conducted with and without an excitation field of 2 kV/mm, and with and without shearing the fluid at shear rates up to 9500 s^{-1}. The condition of the fluid was assessed periodically by measuring the torque response of the clutch to a step application of voltage in respect of both magnitude and speed of response.

Results at the two pole 50 Hz synchronous speed of 3000 rpm indicated that the particles in the fluid were centrifuged over the prolonged test periods. The application of a voltage across the fluid had a negligible effect on this particle migration. The effect of particle migration due to centrifugal and electro-static effects indicate future development requirements for these smart materials.

INTRODUCTION

THE basic concept of using electro-rheological (ER) fluids to develop devices such as clutches have been proposed on many occasions (Brooks, 1982; Stevens et al., 1988). Recent work on the use of ER fluids to construct controllable devices such as clutches and catches (Bullough et al., 1991, 1992a) is showing considerable promise. The development of these devices suffers from particle stability problems associated with ER fluids and demonstrates the need for their continuing development. For example, the development of a high speed ER clutch (Bullough et al., 1991, 1992a, 1992b; Firoozian et al., 1991a, 1991b; Hosseini-Sianaki et al., 1991; Johnson et al., 1991) has highlighted the areas of ER fluid performance and durability under severe operational conditions which are not normally found in laboratory tests. These include rotational speeds of up to 3000 rpm leading to maximum shear rates of 19,000 s^{-1} and maximum centripetal accelerations of 3000 m/s^2. Further, the development has included tests that were conducted over prolonged periods, an important consideration for practical devices.

The basic configuration of the high speed ER clutch is shown in Figure 1, and is essentially a high torque/inertia ratio geometry. Investigations by Firoozian et al. (1991a) have shown that when the ER fluid is energised, torque is transmitted from the rotating outer electrode to the output rotor almost instantaneously. The output rotor needs to have a low inertia so that the speed of response of the clutch will be as fast as possible. The rotor inertia can be made low by using a lightweight nonmetallic material which carries only a thin conducting film on its outer surface. It should be emphasised that the outer electrode is connected to the driving motor, which is of high inertia and running at constant speed (Bullough et al., 1991); its inertia is therefore of no significance to the speed of response of the clutch. A practical embodiment of this concept is given in the prototype ER clutch shown in Figure 2. This has a nominal capacity of 2 Nm and has been detailed elsewhere (Bullough et al., 1991, 1992b; Johnson et al., 1991). The ER fluid used is a mixture of lithium polymethacrylate salt moistened by water and carried in a dielectric oil (Stangroom, 1980). The detailed properties were available from earlier tests (Firoozian et al., 1990); in particular, the maximum controllable yield stress is 2–3 kPa at 2.4 kV/mm. For a nominal 2 Nm clutch, the diameter of the rotor is 60 mm, its length is 60 mm and the ER fluid gap is 0.5 mm. Thus, at the typical excitation voltage of 1000 V, the electric field strength across the ER fluid is 2 kV/mm. The rotor inertia is 4×10^{-4} kgm^2, giving a torque/inertia ratio of 0.5×10^4 Nm/kgm^2. However, in the prototype, most of this inertia is in the comparatively thick stainless steel sleeve used for the high voltage electrode on the rotor. Known design changes could reduce this inertia to 0.4×10^{-4} kgm^2 and with the use of a higher 10 kPa controllable yield stress fluid (Johnson et al., 1991; Brooks, 1989), would increase the torque/inertia ratio to 2.5×10^5 Nm/kgm^2.

Previous tests on the torque transmission of this clutch (Bullough et al., 1991, 1992a) had indicated that the transmitted ER torque T_e fell during the several months of the test programme. The reason for this was not clear but might have been caused by fluid degradation due to centrifuging, excessive shearing, dielectrophoresis, or fluid interaction with the materials used to construct the prototype clutch. Indications from the initial test programme were not conclusive. They included a slight increase in pressure within the clutch, and—on stripping the clutch—a thin smeary film was apparent on the inner negative high voltage electrode.

*Author to whom correspondence should be addressed

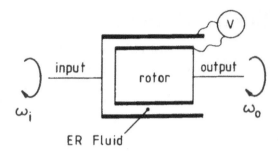

Figure 1. Concentric cylinder ER clutch configuration.

An attempt to evaluate the causes for this performance fall was planned and would include chemical and magnetic resonance imaging of the fluid to assess any changes in the fluid (for example, loss of water content). A prolonged series of endurance tests was also planned to include a systematic investigation of possible centrifugal effects on the fluid. The results of these centrifugal tests are presented in this article.

TEST FACILITY

A schematic layout of the test facility is shown in Figure 3. An important conceptual ideal behind this high speed ER clutch was that it should latch the load onto a constant speed drive. In order to achieve this, an oversized (8 kW) speed controllable (0–3000 rpm) dc electric motor (1) was used. The effects on motor speed of clutching in typical loads was shown to be negligible. The clutch input member (2), supported in two rolling element bearings (3), was connected to

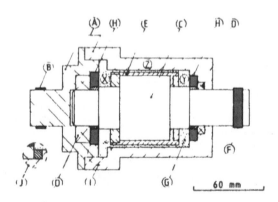

Ⓐ Input member (stainless steel)
Ⓑ Earthing and thermocouple slip-rings
Ⓒ Output member (glass reinforced nylon)
Ⓓ Rolling element bearing
Ⓔ High voltage electrode (stainless steel)
Ⓕ High voltage slip-ring
Ⓖ ER fluid
Ⓗ Viton lip seals
Ⓘ Annular cavity
Ⓙ Annular cavity blocked with nylon insert
Ⓧ, Ⓨ, Ⓩ thermocouple positions

Figure 2. Cross section of prototype ER clutch.

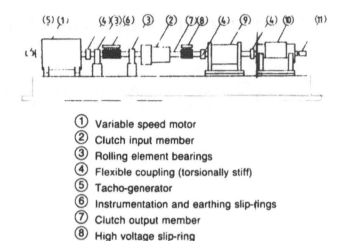

① Variable speed motor
② Clutch input member
③ Rolling element bearings
④ Flexible coupling (torsionally stiff)
⑤ Tacho-generator
⑥ Instrumentation and earthing slip-rings
⑦ Clutch output member
⑧ High voltage slip-ring
⑨ Dynamic torque transducer
⑩ Powder brake dynamometer
⑪ Tacho-generator

Figure 3. Schematic view of experimental test facility.

the motor by a torsionally stiff flexible coupling (4) to eliminate unwanted dynamic effects. The speed of the motor and clutch input member was measured with a small tacho-generator (5). High quality silver/silver-graphite slip-rings and brushes (6) were attached to the input member and enabled the clutch outer casing to be earthed. The signals from the thermocouples measuring the fluid temperatures were also transmitted by these slip-rings to the stationary measuring instrumentation.

The output member of the clutch (7) carried a further set of silver/silver-graphite slip-rings and brushes (8); these were used to transmit the high voltage excitation and associated current to the inner electrode. In these tests, this inner electrode high voltage was of negative potential, the outer electrode being earthed. The high voltage power supply (Wallis R23-50N-37) provided the clutch excitation and was capable of generating step voltages of 2 kV, with negligible voltage overshoot and negligible time delay, at a maximum current of 50 mA.

The output torque of the clutch was measured using the 0–4 Nm dynamic torque transducer (9) (Vibrometer TG/2). Unfortunately, for a torque transducer to be sensitive, it is necessary for it to be torsionally flexible. In this case, this meant that the torque transducer had a natural frequency of 560 Hz, and when connected to the clutch rotor, the resulting natural frequency was 110 Hz, well within the operating torque frequency bandwidth of the clutch. Nonetheless, the actual response could be estimated using identification techniques and digital simulation software (Firoozian et al., 1991a). The accuracy of this technique had been confirmed on a specially stiff torque transducer (Hosseini-Sianaki et al., 1992). Various loadings could be applied in a constant, or quasi-constant manner, by a 0–10 Nm powder brake dynamometer (10) (Vibrometer 2PB 43), which had a fairly slow time constant of about 200 ms. This dynamometer could be used to hold the output of the clutch stationary, or apply a constant or slowly varying torque while it was rotating. The rotational speed of the output member was measured with a small tacho-generator (11).

The measurements required were the clutch excitation voltage and current, input and output speed of the clutch, transmitted clutch torque, and the fluid temperatures. Each measurement would be required simultaneously and, in view of the fast response, at a fast sampling rate in order to use the results for identification studies. This was achieved by using a high speed computer data logging system (Micro-link 4000) which could capture the six channels simultaneously—to 12 bit resolution—at sampling rates of up to 1 MHz.

TEST PROCEDURES AND RESULTS

In order to investigate possible fluid deterioration due to centrifuging, a quick and easily repeatable standard test was required. Previous work (Firoozian et al., 1991a) indicated that the response of the transmitted torque to a step increase in excitation voltage was suitable. This test quickly gives values of the electronic to shear stress time delay, t^*, and the electro-rheological torque, T_e, transmitted by the clutch. This standard step test was therefore chosen, and is now detailed.

Standard Step Test

Fluid temperature is an important parameter and was monitored by the thermocouples incorporated in the prototype clutch. For the standard step test, a standard value of 28°C was chosen. To achieve this temperature, the outer member of the clutch was gently heated with a heat air gun; in order to ensure uniform heating, the clutch was rotated at 100 rpm. This speed was very low compared to the 3000 rpm used in the centrifuging tests to avoid significant unlogged centrifuging. Seal friction caused the unconstrained output member to rotate at the same speed as the input member, namely 100 rpm; there was therefore no fluid shearing and hence no internal heating of the fluid. On the infrequent occasions when the fluid needed to be cooled, this was easily achieved by letting the clutch cool naturally. When the fluid temperature was 28°C, the motor, and hence the input member of the clutch, was spun at 667 rpm while the output member was held stationary by clamping the output of the torque transducer. The combined torque due to fluid viscosity and slip-ring seal friction (i.e., $T_o + T_f$) was then measured using the torque transducer. A step voltage of 1000 V, giving an excitation field strength of 2 kV mm, was then applied for 0.8 seconds to the inner electrode, and the total transmitted torque measured (i.e., $T_\Sigma = T_e + T_o + T_f$). The motor and excitation voltage were then immediately switched off to minimise shearing, and hence heating, of the fluid. Typically, the test was completed in under 10 seconds once the standard temperature of 28 C had been reached. The results were analysed as detailed in Firoozian et al. (1991a) to obtain t^* and T_e; these are defined on the typical result shown in Figure 4.

Test A—Centrifuge at Zero Volts

This test was designed to investigate any centrifuging effects at the operating speed of 3000 rpm with no excitation voltage applied to the ER fluid. Further, there was no shear-

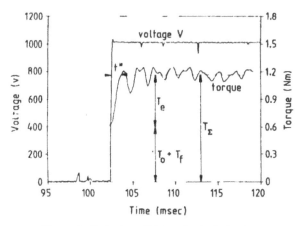

Figure 4. *Typical result from standard step test.*

ing of the fluid; that is, the input and output members were both rotating at 3000 rpm, with seal friction sufficient to prevent any slippage.

In preparation for the test, the clutch was carefully cleaned and filled with fresh fluid, and the bottle of fluid was thoroughly shaken to ensure that the fluid was homogeneous. Suitable venting of the clutch enabled all air to be excluded from the clutch assembly; this ensured that it was completely filled with fluid. An initial series of three standard step tests were performed, and the resulting values for $T_o + T_f$, T_e and t^* were found to be repeatable. The clutch was then rotated at 3000 rpm for one minute, followed by two standard step tests, again used to check for repeatability. The sequence of centrifuging followed by standard step tests was then repeated. The length of time of centrifuging between standard step tests was, however, gradually increased until fifteen minutes of total accumulated centrifuging time had elapsed. The centrifuging time between standard step tests was then kept constant at ten minutes for the remainder of the 720 minutes total centrifuging time of the test. Since it was not practical to complete the test in a working period, it was necessary to incorporate overnight stops. During the overnight stops, which were typically of sixteen hours duration, the clutch was stationary. The timings of these overnight stops are shown on the figures. At the start of the following working period, two standard step tests were performed, followed by the normal centrifuging/standard step test cycle. Frequently, the first few tests indicated a lower value for T_e, with the value rising in a few test cycles to a similar value to that at the end of the previous working period. This effect is clearly shown in Figure 5, which shows the values of $T_o + T_f$ and T_e for this test. The rapid falls in $T_o + T_f$ are probably due to the reduction of seal friction after a short running duration and also the well-known short time thixotropic effects often associated with ER fluids.

Clearly there is an effect of centrifuging, with T_e falling as the accumulated centrifuging time increases. This was further confirmed when the clutch was stripped down at the end of the test and the self-supporting, but easily deformed paste of particles (shown in Figure 6) was found in the annular cavity shown in Figure 2 detail (1). The ER fluid was also less viscous due to the lower number of particles in it.

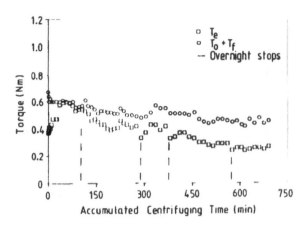

Figure 5. *Result from test A—centrifuging at 3000 rpm with zero excitation voltage, annular cavity unblocked.*

This indicated that the particles were being centrifuged into the annular cavity. The observed fall in ER effect occurs due to this reduced particle content of the fluid. When the annulus becomes full, it could be argued that the concentration would not fall further and that the ER effect would tend to stabilise. This is precisely what is seen in Figure 5, where the value of T_e is reaching an asymptotic value of about 0.26 Nm towards the end of the test period.

The centrifuging of particles may also indicate the reason for the apparently curious increase in T_e over the first 35 minutes of the test. Consideration of Figure 2 indicates that particles in the region Y must migrate through the ER fluid gap before reaching the annular cavity. Thus, in the early stages, the concentration of particles in the gap may increase, providing a possible explanation for this initial rise in T_e.

A further interesting feature to emerge from this test was that the electronic to shear stress time delay t^* was essentially constant at about one millisecond. This indicates that particle concentration has little effect on this important ER fluid property. Examination of the stripped down clutch did not show any adverse reactions between the ER fluid and the

materials used in the construction of the prototype, namely stainless steel, glass reinforced nylon, viton seals, and copper/constantin thermocouples. Any chemical degradation of the fluid from unsuitable constructional materials therefore appears unlikely.

Test B—Centrifuge at 1000 Volts

This test was designed to investigate whether applying an excitation voltage of 1000 volts across the ER fluid—which gives an electric field strength of 2 kV/mm—while centrifuging at 3000 rpm would affect the results found from test A. That is, whether the application of an excitation field would prevent, or possibly enhance, the centrifuging of the particles, and thus the change in the transmitted ER torque T_e. The test procedure was identical to test A, again starting with fresh fluid, but with the excitation voltage of 1000 V applied during the centrifuging period. During overnight stops, the clutch was stationary and no excitation voltage was applied.

The results for this test are shown in Figure 7; again, T_e rises to a maximum of 0.6 Nm followed by a fall to a lower level of 0.27 Nm at the end of the test. On stripping down the clutch at the end of the test, a similar paste of particles was again found in the annular cavity. The results, when compared with those of test A, indicate that applying an excitation voltage during centrifuging has no discernible effect in preventing or enhancing the migration of the particles to the annular cavity.

Test C—Test A but with Annular Reservoir Blocked

This test was identical to test A but the annular cavity was blocked with nylon as shown in Figure 2 detail (J). This was done in order to investigate the effects of not allowing significant quantities of particles to centrifuge out of the fluid into the annular cavity. The annular cavity had been originally incorporated to provide a fluid reservoir to ensure that the clutch gap would always be full of fluid even in the event of a slight leakage. Other geometric configurations may eliminate the need for this annular cavity (for example, if the clutch axis was vertical).

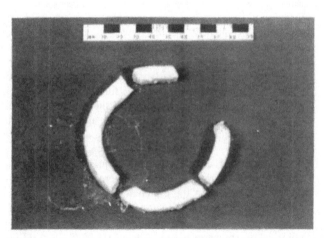

Figure 6. *Particle paste found in annular cavity after conclusion of test A.*

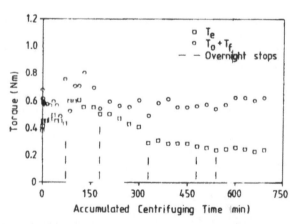

Figure 7. *Result from test B—centrifuging at 3000 rpm with excitation voltage of 1000 V (electric field strength of 2 kV/mm), annular cavity unblocked.*

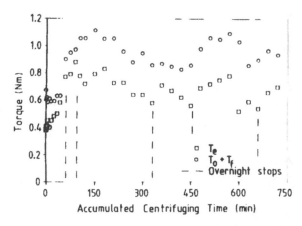

Figure 8. *Result from test C—centrifuging at 3000 rpm with zero excitation voltage, annular cavity blocked.*

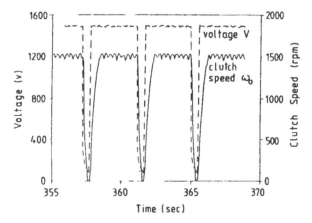

Figure 9. *Excitation voltage cycle and resulting output member speed changes for test D, clutch input speed of 1500 rpm, maximum fluid shearing rate of 9500 s⁻¹.*

The results are shown in Figure 8. Again, the initial values of $T_o + T_f$ and T_e are similar, indicating that the initial fluid condition was the same. The value of T_e again rises rapidly, but this time to a higher level of about 0.8 Nm and then fluctuates significantly. This may be due to a concentration of solid particles in the gap, perhaps at times becoming too concentrated and causing a reduction in ER effect. The higher levels of $T_o + T_f$ may also be due to this packing of particles. The standard step test involves some fluid shearing, which may serve to disrupt this overpacking and hence give rise to the large scattering of the results.

On stripping the clutch down, the fluid in the gap was more viscous than fresh fluid, indicating that the particle concentration had increased. Further, there were also some similar pasty deposits in the outer corners of the regions X and Y in Figure 2, again indicating the centrifuging of fluid particles. Further, a thin smeary deposit was again apparent on the inner negative high voltage electrode. Interestingly, the value of t^* was the same as in tests A and B, again indicating that this ER fluid property does not significantly depend on the particle concentration.

Test D—Realistic On/Off Control Test

From previous work (Bullough et al., 1992b; Johnson et al., 1991; Firoozian et al., 1991b), it has become apparent that due to the time dependent hysteresis in ER fluids, the best control strategy is probably to use pulse-width modulation techniques; that is, pulse the voltage on and off. This test was therefore designed to investigate centrifuging under these more realistic operating conditions. For this test, the powder brake dynamometer was used and was set to a torque which was just below the total torque transmitted by the clutch (i.e., $T_\Sigma = T_e + T_o + T_f$). Thus, when the clutch is not energised, the dynamometer holds the output member stationary. When the clutch is energised, the dynamometer applies the torque, but the clutch quickly accelerates the dynamometer rotor until no slipping occurs. This ON/OFF voltage cycle and resulting clutch output member speed is shown in Figure 9 for the test conducted. In order to prevent excessive heating due to shearing of the fluid, the speed was limited to 1500 rpm and the pulsing was of two

and a half minutes duration between standard step tests. This time period of pulsing was gradually increased to nine minutes, and was then held at this time for the remainder of the test. It should be noted that for this test, the annular cavity was kept blocked.

The results are shown in Figure 10. Again, the initial values of $T_o + T_f$ and T_e are similar to all previous tests, confirming that the initial condition of the fluid was similar. However, this time the value of T_e increased more quickly to a lower peak of 0.47 Nm before dropping away much quicker to the lower level of 0.26 Nm. It should be emphasised that for this test, the rotational speed was 1500 rpm, giving a centripetal acceleration of 750 m/s² that is one quarter of the value at 3000 rpm used for the other tests. Clearly then, the shear rate of 9500 s⁻¹ in the fluid is contributing to the deterioration of the ER effect. This feature is currently being investigated further.

DISCUSSION

The results indicate that centrifugal effects are a significant factor in the deterioration of the ER effect in the clutch

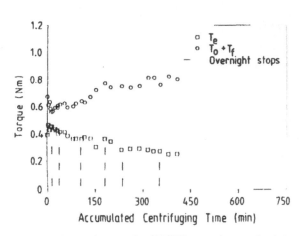

Figure 10. *Result for test D—ON/OFF control test, clutch input speed of 1500 rpm, annular cavity blocked.*

application when operating at a realistic engineering speed of 3000 rpm. It should be emphasised that at this speed, the particles at the 30 mm radius surface of the inner electrode are experiencing a centripetal acceleration of approximately 3000 m/s² (300 times greater than gravity). This indicates that separation tests during fluid development need to be conducted under these severe conditions and not only gravitational loading. Whether a neutral density ER fluid— one where fluid and particle density are identical—would eliminate the problem is arguable. Certainly, the results draw attention to problems for the colloidal chemist/rheologist to investigate.

The results indicate that it is the decrease in particle concentration that causes the reduction in the yield stress; the use of surfactants to overcome the problem should be given a high priority. Other areas that certainly need further investigation are changing the gap size between the inner and outer electrodes, the reversal of field direction across the ER fluid in the clutch (although there seems little scope for significant improvement in reducing particle migration due to this cause), and the effect of fluid shearing on deterioration of the ER effect.

CONCLUSIONS

Results from an ER clutch operating at the two pole 50 Hz synchronous speed of 3000 rpm indicate that the particles in the fluid are centrifuged out too rapidly for the more prolonged duty cycles. The application of a voltage across the ER fluid has a negligible effect on the particle migration.

Depending on the geometric details, this migration of particles can lead to a significant variation of particle density in the fluid, thus altering the yield stress and the viscosity of the ER fluid. Interestingly, the electronic to shear delay t^* is unaltered by this change in particle density.

The results of this investigation clearly and urgently draw the attention of fluid developers to the necessity for adequate treatment of this problem.

NOMENCLATURE

E = excitation field strength (V/mm)
t^* = electronic to shear stress time delay (ms)
T_e = electro-rheological torque (Nm)
T_f = friction torque (Nm)
T_o = viscous torque (Nm)
T_Σ = total clutch torque $(T_o + T_e + T_f)$

V = excitation voltage
ω_i = angular speed of input member (rpm)
ω_o = angular speed of output member (rpm)

ACKNOWLEDGEMENTS

The authors are grateful to the Science and Engineering Research Council for their support of this work through the Specially Promoted Programme: The Design of High Speed Machines. Thanks are also due to Mr. D. Butcher who constructed the prototype clutch and test facility.

REFERENCES

Brooks, D. A. 1982. "Electro-Rheological Devices", *Chartered Mechanical Engineer* (Sept.):91–93.

Brooks, D. A. 1989. "Fluids Get Tough", *Physics World*, 2:35–38.

Bullough, W. A., et al. 1991. "The Electro-Rheological Catch/Latch/Clutch", *Implementation and Prospects, Proc. I. Mech. E., Eurotech Direct '91 Conf. ICC, Birmingham*, Paper C414/070, pp. 129–136.

Bullough, W. A., et al. 1992a. "Quick-Acting Drive Connections", UK Patent Application No. GB 2,252,148 A, Publ. 29.7.92

Bullough, W. A., et al. 1992b. "Fast Pick Up and Drop Load Performance of a Low Electrical and Mechanical Time Constant Electro-Rheologically Based Clutch", *Mechatronic Systems Engineering, Vol. 4*. Dordrecht, Netherlands: Kluwer Academic Publishers, pp. 315–327.

Firoozian, R., D. J. Peel and W. A. Bullough. 1990. "Magnetic Effects in an Electro-Rheological Controller", *Proc. I. Mech. E., Int. Conf. Mechatronics, Designing Intelligent Machines. Cambridge*, pp. 231–238.

Firoozian, R., et al. 1991a. "Dynamic Torque Response of an Electro-Rheological Clutch", *DE 35 Rotating Machinery and Vehicle Dynamics, Miami*, ASME, pp. 295–301.

Firoozian, R., et al. 1991b. "Control Techniques for Electro-Rheological Fluids." *Proc. 24th ISATA Int. Symp. on Automotive Technology and Automation, Florence*, pp. 479–487.

Hosseini-Sianaki, A., et al. 1991. "Operational Considerations in the Use of an Electro-Rheological Clutch Device", *Proc. China Fluid Power Society, 1st Fluid Power Trans. and Control Symposium, Beijing*, pp. 591–595.

Hosseini-Sianaki, A., et al. 1992. "Experimental Measurements of the Dynamic Torque Response of an ERF in the Shear Mode", *Proc. E. R Conf., Carbondale, IL, October, 1991*, pp. 219–235.

Johnson, A. R., et al. 1991. "Electro-Rheological Clutch under Inertial Loading", *Proc. MPT 91, Int. Conf. of Motion and Power Transmissions, JSME, Hiroshima*, pp. 1016–1021

Stangroom, J. E. 1980. "Improvements (Further Improvements) in or Relating to Electric Field Responsive Fluids". UK patent 1,570,234.

Stevens, N. G., et al. 1988. "An Experimental Study of Electro-Rheological Torque Transmission", *Trans. ASME, Journal of Mechanisms, Transmissions and Automation in Design*, 110:182–188.

Decomposition of the Pressure Response in an ER Valve Control System

M. WHITTLE, R. FIROOZIAN, D. J. PEEL AND W. A. BULLOUGH*

Department of Mechanical and Process Engineering
The University of Sheffield
P.O. Box 600
Sheffield S1 4DU, United Kingdom

ABSTRACT: Experimental data from tests on an ER valve pertinent to the development of a controller for high speed machine duty has been analysed to show that three common, underlying modes of response are present. This is demonstrated for a range of industrial scale flow velocities and electrode dimensions, in the time and frequency domains. The dependence of steady-state pressure on the electric field is also discussed.

INTRODUCTION

THE mechanisms involved in the generation of field-induced ER stress remain the subject of some controversy. Several computer-based multi-particle models have been advanced (Whittle, 1990; Klingenberg and Zukoski, 1990; Bonnecaze and Brady, 1992) and some experimental results have been reported (Tao, 1991), but often these do not cover the range of engineering requirements—an important reason for the continued development of ER fluids. Sometimes theories are presented for an unrealistic fluid or zone of operation.

In the present work, high quality results from an engineering scale test programme have been analysed for a typical good fluid. This has been done with two aims in mind. Initially, the time constants and associated data are required for control engineering considerations. Secondly, and more relevant to the present work, the same general underlying forms of response may be useful to rheologists, physicists and chemists inasmuch as they limit the number and types of contributory mechanisms of shear stress or pressure generation involved. In the present work, it has been possible to show that the step (short term time domain) and frequency responses of pressure drop in a valve subject to voltage can both be modelled by combining three basic underlying responses. In addition, the non-linear nature of the steady-state (long term time domain) pressure as a function of field strength is described. The investigation covers the presently envisaged range of practical field strengths 0–2.5 MVm^{-1} over an electrode separation 0.5–1.0 mm and fluid velocities 0.14–1.4 ms^{-1} (approximate maximum shear rate 25,000 s^{-1} according to the Bingham model) for one well-documented ER fluid. This fluid is based on lipol (lithium polymethacrylate) at a volume fraction of 30% containing 10–15% water with a nominal yield stress $\tau_e > 3$ to 4 kPa at ~4 MVm^{-1} and particle size averaging 10 to 15 microns.

Each of the five valves investigated consists of two near parallel channels of mean radius 28.8 mm. These and the test apparatus, which involves high speed quality instruments and data loggers, has been described elsewhere (Bullough and Peel, 1990). For ER fluids, the flow mode has advantages over other systems of experimentation by virtue of the ease of transient pressure measurement (compared with the torque in a rotating system) and the ability to control the crucially important temperature of the fluid. The penalty is a variation of shear stress over the electrode gap.

In making these results available to rheologists, physicists and chemists, it is one of the aims of the authors to stimulate collaborative work towards the establishment of the modus operandi of ER fluids and the development of the properties required for their industrial application. It is hoped that the results may be of use to researchers following alternative approaches.

STEADY-STATE PRESSURE CHARACTERISTICS AND RAMP RESPONSE

The short time (0–10 ms) response to a step function voltage input will be described in the next section. At longer times, the current and pressure response to this input normally reach a steady state. The approach to this plateau may contain some very slow relaxations indeed, particularly in the current response (Whittle et al., 1991). However, the pressure has normally reached its limiting value after 0.5 s and we have therefore defined "steady state" for our purposes to be at 0.75 s.

Electrorheological properties in general display markedly non-linear characteristics when plotted against the field.

---

*Author to whom correspondence should be addressed

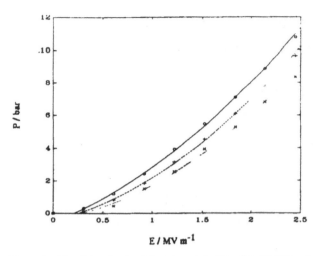

Figure 1. *Fits of the steady-state pressure by Equation (1). Valve A,* —C— *Q = 3 Lit min⁻¹;* —+— *Q = 9 Lit min⁻¹.* —x— *Q = 15 Lit min⁻¹.*

Here, we concentrate on the pressure response. As shown in Figure 1, we find that the steady-state pressure P_{ss} developed at field E is described extremely well over the domain of test values herein by an empirical expression involving a threshold field strength E_0 with the form:

$$P_{ss} = p_1(E - E_0) + p_2(E - E_0)^2 \quad ; \quad E > E_0$$
$$P_{ss} = 0 \qquad\qquad\qquad\qquad ; \quad E < E_0 \tag{1}$$

where p_1 and p_2 are coefficients. A theoretical expression for the pressure (Bullough and Peel, 1990) suggests that it should be proportional to the valve length l, and since we are interested in comparing valves of different dimensions, we define the normalised coefficients:

$$p_1^* = p_1/l \quad ; \quad p_2^* = p_2/l \tag{2}$$

Valve dimensions used in this work are given in Table 1.

It is also appropriate to compare data with respect to a mean fluid velocity U rather than the volume flow rate Q, where

$$U = Ql/Ah \tag{3}$$

for total electrode surface area A, length l and gap h. Param-

Table 1. *Valve dimensions: length l, gap h, surface area A.*

Valve	l (mm)	h (mm)	A (m²)
A	100	0.5	.0362
B	50	0.5	.0181
C	100	1.0	.0362
D	140	0.5	.0507
E	100	0.75	.0362

Table 2. *Parameters obtained from "steady-state" results fitted by Equation (1).*

Valve	Q (Lit min⁻¹)	U (m s⁻¹)	p_1^* (Bar MV⁻¹)	p_2^* (Bar m MV⁻²)	E_0 (MVm⁻¹)
A	3.0	0.276	28.8	8.97	0.208
	9.0	0.828	21.6	10.6	0.255
	15.0	1.38	16.7	11.0	0.292
B	3.0	0.276	42.4	14.5	0.263
	9.0	0.828	36.4	29.0	0.468
	15.0	1.38	0.0	35.6	0.375
C	3.0	0.138	11.1	8.23	0.141
	9.0	0.414	13.0	9.72	0.179
	15.0	0.69	12.5	10.4	0.209
D	3.0	0.276	33.2	9.88	0.222
	9.0	0.828	22.5	14.4	0.25
	15.0	1.38	26.4	12.8	0.46
E	3.0	0.184	26.2	6.3	0.202
	9.0	0.552	24.9	12.3	0.25
	15.0	0.92	27.7	14.4	0.422

eters obtained by fitting Equation (2) to this data are shown in Table 2.

Data for a ramped voltage input contains more finely resolved information at low fields, and an example is shown in Figure 2 to demonstrate that the fit is extremely good over the whole range of interest. Small deviations occur only at relatively high fields. The parameters obtained are comparable with those for the steady state.

Plotted against the mean flow rate in Figure 3, the values of E_0 suggest a roughly linear relationship with a non-zero intercept. The threshold values are similar to those we have recently reported (Whittle et al., 1991), derived from a related study of the steady-state current density i where an empirical expression involving only a quadratic term was found to apply:

$$i = \Xi^*(E - E_0)^2 \quad ; \quad E > E_0 \tag{4}$$

where Ξ^* is a coefficient. If the threshold values for pressure and current are taken as roughly equivalent, a simple empirical relationship then follows between the pressure and current:

$$P \sim p_1(i/\Xi^*)^{1/2} + p_2(i/\Xi^*) \tag{5}$$

PRESSURE RESPONSE TO VOLTAGE STEP INPUT

After removal of the sinusoidal pump noise, the pressure response to a voltage step under the conditions of this experiment is qualitatively characterised by a short delay of ~0.2 ms followed by a rise over ~4 ms, usually with some oscillatory behaviour, which gives the appearance of an overshoot. Beyond 10 ms, there is a slow increase to the steady-state value, which is normally reached after ~0.5 s.

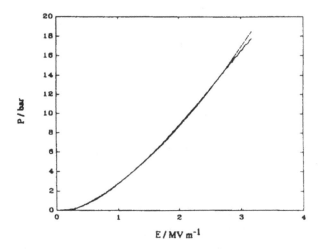

Figure 2. *Fit of Equation (1) to ramp data. Valve A, Q = 9 Lit min⁻¹, ramp 1.0 MVm⁻¹s⁻¹: —— experiment; ----- fit.*

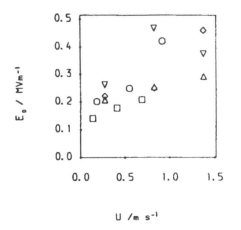

Figure 3. *Threshold voltage E₀ plotted against mean flow velocity U. —△— Valve A; —   — Valve B; —□— Valve C; —   — Valve D; —   — Valve E.*

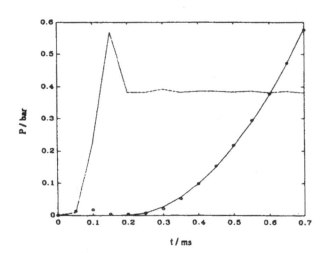

Figure 4. *The early pressure response described by a quadratic. Valve A, Q = 3 Lit min⁻¹, voltage step = 600 V. ○ Experimental data; —— quadratic fit with delay; ----- voltage input on an arbitrary scale.*

The early part of the response is shown for a typical case in Figure 4, where it is compared with the voltage step profile. On this timescale, the pressure response $P(t)$ can be described by a delay t_0 followed by a quadratic rise:

$$P(t) = \alpha(t - t_0)^2 \quad ; \quad t > t_0 \tag{6}$$

where α is a coefficient. A theoretical prediction of the response for this system via the time dependent Navier-Stokes equation is not currently available. However, since a step function has been used as the input, numerical differentiation of the experimental response gives directly a trace corresponding to the transfer function for the system. This yields a form that suggests a sum of damped oscillatory terms T_{2n} as a possible decomposition:

$$T_{2n} = \frac{a_{2n}}{\beta_{2n}} \exp(-t'/\tau_{2n}) \sin(\beta_{2n}t') \tag{7}$$

Here, T_{mn} is the nth term of order m, $t' = t - t_0$, and a_{mn} is a coefficient with units Bar s⁻². Henceforth, wherever possible, we drop the subscripts for a_{mn}, β_{mn} and τ_{mn} and adopt the convention that they are implied by the left-hand side of the equation.

These functions are commonly described as the impulse response of a standard second order differential equation (Ogata, 1978) for an arbitrary variable $x(t)$:

$$\frac{d^2x}{dt^2} + 2\zeta\omega_n \frac{dx}{dt} + \omega_n^2 x = a\delta(0) \tag{8}$$

with a solution $X(s)$ in Laplace transform space $(dx(0)/dt = 0$ and $x(0) = 0)$

$$X(s) = \frac{a}{s^2 + 2\zeta\omega_n s + \omega_n^2} \tag{9}$$

where ζ is the damping ratio and ω_n is the natural frequency. For critically damped systems, $\zeta = 1$. In terms of Equation (7), we can identify

$$\omega_n = \sqrt{\frac{1}{\tau^2} + \beta^2} \quad \text{and} \quad \zeta = \frac{1}{\sqrt{1 + \tau^2\beta^2}} \tag{10}$$

Numerical differentiation amplifies signal noise, and it is better to fit the integrated form directly to the experimental data,

$$P_{2n}(t) = \int_0^t T_{2n}(\beta, \tau, t)dt \tag{11}$$

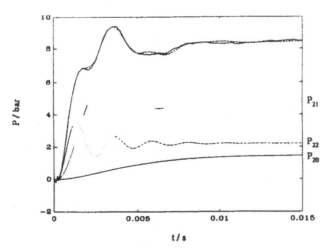

Figure 5. *The pressure response to a step function voltage input and its resolution into three terms. Valve A, Q = 9 Lit min⁻¹, field step = 2.4 MVm⁻¹. The upper traces compare the sum of the three terms P_{20}, P_{21}, and P_{22} (——) with experiment (——).*

giving

$$P_{2n} = \frac{a\tau^2}{1 + \beta^2\tau^2}\left[1 - \exp(-t'/\tau)\right.$$

$$\left. \times \left(\frac{1}{\beta\tau}\sin\beta t' + \cos\beta t'\right)\right] \qquad (12)$$

In fact, we find that a good description of the data is obtained if three second order terms are used, one of which (identified as the "zeroth" $n = 0$ term, P_{20}) is critically damped. In practice, at intermediate and long timescales, this term is almost indistinguishable from a first order response; in integrated form,

$$P_{11} = a\tau[1 - \exp(-t'/\tau)] \qquad (13)$$

where a_{11} has units Bar s⁻¹. However, as we have seen, the very early (<0.5 ms) part of the pressure response is quadratic rather than linear. It may be easily verified that this is characteristic of second order terms by expanding the above expressions for small t. Our model for the step response can now be summarised as

$$P(t) = P_{20}(t') + P_{21}(t') + P_{22}(t') \qquad (14)$$

with a corresponding transfer function

$$T(t) = T_{20}(t') + T_{21}(t') + T_{22}(t') \qquad (15)$$

The resulting function has been fitted to the data for all valves using a Nelder-Mead simplex algorithm (Press et al., 1986), and an example is shown in Figure 5. In most cases, the fit is extremely good, although in one or two cases, small discrepancies arise, which we suspect are due to non-linearities in the response. The parameters obtained are collected in Tables 3a and 3b. We quote normalised coefficients for the integrated forms A_{mn}^* and the effective long time pressure $P_\infty \sim P_{ss}$ obtained by fitting

$$P(t) = P_\infty[A_{20}^* P_{20}^*(t') + A_{21}^* P_{21}^*(t') + A_{22}^* P_{22}^*(t')] \qquad (16)$$

where

$$P_{2n}^*(t') = P_{2n}(t')\frac{(1 + \beta^2\tau^2)}{a\tau^2} \quad \text{and} \quad A_{2n}^* = \frac{a\tau^2}{P_\infty(1 + \beta^2\tau^2)} \qquad (17)$$

introduce coefficients A_{2n}^*, which sum to unity and effectively give the proportions of each component contributing to the step response.

Table 3a. *Parameters obtained from the step function response, E = 1.2 MVm⁻¹.*
(S.D. is the standard deviation of the fit.)

Valve	Q (Lit min⁻¹)	P_∞ (Bar)	A_{20}^*	ω_{20} (Hz)	A_{21}^*	ω_{21} (Hz)	ζ_{21}	A_{22}^*	ω_{22} (Hz)	ζ_{22}	t_0 (ms)	S.D.
A	3.0	3.25	0.445	114.5	0.391	164.0	0.397	0.163	465.7	0.278	0.212	0.022
	9.0	2.16	0.424	129.4	0.295	177.2	0.306	0.218	538.0	0.262	0.31	0.032
	15.0	2.16	0.236	60.3	0.594	194.5	0.65	0.170	481.7	0.141	0.191	0.065
B	3.0	3.85	0.591	114.7	0.247	174.7	0.359	0.162	552.0	0.257	0.294	0.023
	9.0	2.71	0.424	129.4	0.223	195.8	0.277	0.292	698.0	0.268	0.29	0.085
	15.0	1.69	0.382	95.3	0.416	188.7	0.398	0.202	457.4	0.212	0.03	0.106
C	3.0	2.62	0.013	76.9	0.928	155.3	0.78	0.059	457.3	0.208	0.208	0.021
	9.0	3.65	0.024	79.8	0.821	187.7	0.764	0.154	604.3	0.261	0.242	0.086
	15.0	3.88	0.150	106.8	0.614	191.0	0.712	0.237	524.3	0.233	0.199	0.124
D	3.0	4.41	0.411	100.7	0.465	167.4	0.464	0.124	414.1	0.193	0.262	0.018
	9.0	3.54	0.363	86.5	0.442	178.0	0.463	0.196	464.9	0.186	0.222	0.040
	15.0	2.57	0.327	58.7	0.463	201.1	0.582	0.210	529.2	0.165	0.215	0.064
E	3.0	4.89	0.511	138.4	0.352	171.0	0.473	0.137	495.7	0.295	0.299	0.031
	9.0	5.1	0.607	173.1	0.137	189.0	0.243	0.256	643.9	0.317	0.357	0.057
	15.0	4.41	0.235	79.6	0.531	193.9	0.622	0.234	567.5	0.351	0.205	0.129

Table 3b. Parameters obtained from the step function response, $E = 2.4$ MVm^{-1}.
(S.D. is the standard deviation of the fit.)

Valve	Q (Lit min^{-1})	P_∞ (Bar)	A_{20}^*	ω_{20} (Hz)	A_{21}^*	ω_{21} (Hz)	ζ_{21}	A_{22}^*	ω_{22} (Hz)	ζ_{22}	t_0 (ms)	S.D.
A	3.0	8.91	0.372	76.2	0.515	146.5	0.443	0.113	400.9	0.138	0.157	0.056
	9.0	8.45	0.174	67.7	0.565	176.8	0.365	0.261	448.7	0.161	0.189	0.080
	15.0	7.42	0.214	72.7	0.477	178.0	0.344	0.309	507.1	0.178	0.180	0.114
B	3.0	11.26	0.352	87.9	0.482	160.7	0.463	0.166	453.3	0.202	0.205	0.111
	9.0	11.30	0.186	84.7	0.515	187.0	0.343	0.298	541.4	0.228	0.238	0.193
	15.0	9.51	0.182	93.1	0.484	188.6	0.338	0.334	634.4	0.307	0.237	0.178
C	3.0	11.79	0.596	140.8	0.290	169.2	0.436	0.114	427.0	0.210	0.153	0.135
	9.0	13.40	0.329	205.3	0.444	184.6	0.421	0.226	443.3	0.278	0.176	0.214
	15.0	13.88	0.148	220.4	0.530	189.0	0.481	0.322	437.7	0.282	0.192	0.227
D	3.0	12.2	0.573	82.0	0.257	114.0	0.491	0.170	373.0	0.281	0.190	0.067
	9.0	12.35	0.143	52.5	0.594	179.7	0.408	0.263	379.2	0.161	0.239	0.152
	15.0	10.86	0.135	55.5	0.567	185.2	0.409	0.298	424.5	0.192	0.219	0.094
E	3.0	13.84	0.253	70.4	0.607	163.9	0.635	0.139	431.0	0.199	0.221	0.099
	9.0	16.18	0.286	139.6	0.396	188.6	0.378	0.318	425.2	0.225	0.244	0.253
	15.0	15.94	0.173	120.0	0.419	189.5	0.360	0.408	531.2	0.337	0.264	0.220

Table 4a. Parameters obtained from fits in the frequency domain: Valve A, bias $E = 1.2$ MVm^{-1}.

Q (Lit min^{-1})	$Y(0)$ (Bar MV^{-1})	B_{11}^*	τ_{11} (ms)	B_{20}^*	ω_{20} (Hz)	B_{21}^*	ω_{21} (Hz)	ζ_{21}
3.0	59.2	0.31	97.3	0.032	44.2	0.657	165.3	0.584
9.0	50.5	0.477	55.9	0.006	230.6	0.523	203.6	0.724
15.0	44.1	0.522	43.8	0.039	160,000	0.439	215.2	0.747

Table 4b. Parameters obtained from fits in the frequency domain: Valve A, bias $E = 2.4$ MVm^{-1}.

Q (Lit min^{-1})	$Y(0)$ (Bar MV^{-1})	B_{11}^*	τ_{11} (ms)	B_{20}^*	ω_{20} (Hz)	B_{21}^*	ω_{21} (Hz)	ζ_{21}
3.0	62.8	0.346	58.3	0.567	115.3	0.088	104.7	0.103
9.0	74.0	0.563	62.4	0.318	227.4	0.12	131.8	0.365
15.0	67.7	0.542	124.4	0.037	100.1	0.420	217.6	0.938

Table 5. Admittance derived from the steady-state results: Valve A.

Q (Lit min^{-1})	Y_{ss} (Bar MV^{-1} @ 1.2 MVm^{-1})	Y_{ss} (Bar MV^{-1} @ 2.4 MVm^{-1})
3.0	46.6	68.1
9.0	41.6	67.1
15.0	36.6	63.1

RESPONSE TO A BIASED SINUSOIDAL INPUT

The pressure response to a biased sinusoidal voltage input has been obtained for valve A only. A bias of 600 V and 1200 V d.c. was used in conjunction with sinusoidal voltages of amplitude around 50 V. This amplitude is small enough to produce a linear response in the pressure. After removal of the mean bias pressure, the resulting fluctuations ΔP can be described by the superposition of a phase shifted sinusoidal component ϕ_p at the voltage input frequency ω and an additional component, describing the pump noise, of arbitrary phase ϕ_q at the known pump frequency ω_q.

$$\Delta P = V|Y| \sin(\omega t + \phi_I) + \frac{\delta P}{\delta Q} Q_0 \sin(\omega_q t + \phi_q)$$

(18)

With only two frequencies present, a time domain analysis was viable and the Nelder-Mead simplex algorithm (Press et al., 1986) has again been used to fit this form to the pressure output. The resulting fit is extremely good over all frequencies studied (1–200 Hz). Knowing the phase and amplitude of the input signal, the admittance $|Y|$ and phase change $\tan \phi$ can be found at each frequency. Because of the non-linear steady-state characteristics, these are expected to vary with bias voltage.

The response to a step function has been analysed in the previous section in the time domain as a sum of second order terms. The response in the frequency domain for the same model is obtained (Aseltine, 1958) by taking the one-sided Fourier transform of Equation (15)

$$P(j\omega) = \int_0^\infty T(t) \exp(-j\omega t)dt$$

(19)

The admittance $Y(\omega)$ and phase angle ϕ are then given by

$$Y(\omega) = |P(j\omega)|/V \quad ; \quad \tan \phi = \text{Im}[P(j\omega)]/\text{Re}[P(j\omega)]$$

(20)

The term arising from the delay is here neglected (i.e., t' has been replaced with t). This is justified because the delay is shorter than the reciprocal of the highest frequency studied here. It formally introduces a multiplicative factor of $\exp(-j\omega t_0)$, which for small delays is a flat contribution at relatively low frequencies. We thus obtain [either directly from Equation (7) or by substituting $s = j\omega$ in Equation (9)] for a general second order term

$$Y_{2n}(j\omega) = b\frac{\tau^2[(1 + \beta^2\tau^2 - \omega^2\tau^2) - 2j\omega\tau]}{(1 + \beta^2\tau^2 - \omega^2\tau^2)^2 + 4\omega^2\tau^2}$$

(21)

where the b_{mn} have units Bar s^{-2} V^{-1} and are related to the a_{mn}. Frequency dependent data is available in the range

1–200 Hz. Our analysis of the time domain step function response focused on the range 0–20 ms. Beyond this time regime, there is a slow relaxation to the steady state which we include here as a simple first order response characterised by a single time constant τ_{11} [as defined by Equations (12) and (13)] so slow as to have a negligible effect on the form of the very early response. This has the transform

$$Y_{11}(j\omega) = b\frac{\tau - j\omega\tau^2}{1 + \omega^2\tau^2}$$

(22)

Because the high frequency term P_{22} is beyond 200 Hz, the maximum probed by the experiment, we can neglect this in fitting the frequency domain data. Our frequency domain model is then:

$$Y(j\omega) = Y(0)[B_{11}^* Y_{11}^*(j\omega) + B_{20}^* Y_{20}^*(j\omega)$$

$$+ B_{21}^* Y_{21}^*(j\omega)]$$

(23)

where Y_{mn}^* are normalised to unity at zero frequency [compared with Equation (17)] and the B_{mn}^* are coefficients that sum to unity and give the proportions of each component contributing to the response.

$$B_{2n}^* = \frac{b\tau^2}{Y(0)(1 + \beta^2\tau^2)} \quad ; \quad B_{11}^* = b\tau$$

(24)

The real and imaginary parts of this function, $Y' = |Y| \cos \phi$ and $Y'' = |Y| \sin \phi$, have been fit to the experimental counterparts by minimising the separation in the complex plane. An example is shown in Figure 6, and parameters obtained are given in Tables 4a and 4b.

The frequency domain data is rather limited for a fit of so many parameters; nevertheless, values of ω_{21} are quite consistent with those obtained in the step function analysis. Values of the damping ratio ζ_{21} span a wide range for the high field results, but are of the same order as step function results at low field. One result for ω_{20} in Table 4a seems anomalously large and corresponds to an effectively flat contribution to the response. Values of τ_{11} are comparable to those obtained for the slow component of the current response (Whittle et al., 1991). Values of the admittance at zero frequency $Y(0)$ can also be obtained from the differential of Equation (1) and are given as Y_{ss} in Table 5. They vary with bias voltage because of the non-linearity of the steady-state characteristics. The values obtained compare well with those in Tables 4a and 4b. The ratio B_{21}^*/B_{20}^* might be expected to compare with A_{21}^*/A_{20}^* from the step analysis, but there is wide variation here, possibly as a result of the non-linearity.

Replacing $T(t)$ with the experimental response $P(t)$ in Equation (19) the step response data can be transformed into the frequency domain. In Figure 7, the resulting spectral density $S_p(\omega) = P(\omega)P^*(\omega)$ (here, the asterisk indicates the complex conjugate) is compared with the biased sine data

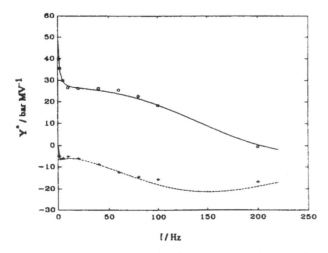

Figure 6. *Fit to the frequency domain data expressed as the admittance $Y(\omega)$. Valve A, $Q = 9$ Lit min^{-1}, bias field $E = 1.2$ MVm^{-1}. Real component of data ○ and fit ——; imaginary component of data + and fit ----.*

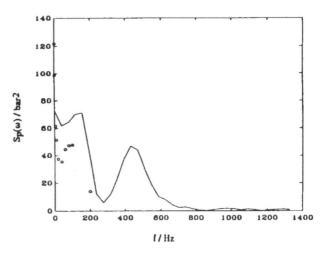

Figure 7. *Spectral density $S_p(\omega)$. — — Biased sine data; —— transformed from step data.*

for one example. Expressed in this form, the two oscillatory components P_{21} and P_{22} are clearly resolved as peaks at ~ 150 and 450 Hz, and the first peak is also apparent in the available biased sine data. The magnitudes are somewhat different as a result of non-linearity in the response to a large step input, while the peak at low frequency in the biased sine data (corresponding to Y_{11} above) is underrepresented by the step data because the lowest frequency accessed in this case is ~ 10 Hz.

CONCLUSIONS

We have found that a model for the pressure response of an ER valve controlled system to a step voltage input can be composed from three second order features, one of which is critically damped. The same model can be transformed into the frequency domain and, with the addition of a slow first order component, describes the response to a biased sinusoidal voltage input. Values of the admittance so obtained can be compared with steady-state data. The steady-state pressure is well described by a second order polynomial with a threshold field strength.

It is to be understood that at this stage of the analysis, no attempt has been made to isolate the effects of wave action in the test rig. In this sense, the results are to be seen as the responses that occur in a valve with its associated system and

are thus authentic from an engineering perspective. Further work will be needed before the response of the ER fluid alone can be determined. This will require a time dependent solution of the equations of motion for fluid in the valve in order to evaluate its response in isolation, and also analyse the response of the test rig to sudden changes in pressure. Nevertheless, the temporal behaviour of this configuration has been shown here to have a regular form, whatever the geometry of the valve and operating conditions. This provides a strong basis for further processing of the data.

REFERENCES

Aseltine, J. A. 1958. *Transform Method in Linear System Analysis*. McGraw-Hill.

Bonnecaze, R. T. and J. F. Brady. 1992. "Dynamic Simulation of an Electrorheological Fluid", *Journal of Chemical Physics*, 96:2183–2202.

Bullough. W. A. and D. J. Peel. 1990. *I. Mech. E.*, C419/043.

Klingenberg. D. J. and C. F. Zukoski. 1990. *Langmuir*, 6:15.

Press, W. H., B. P. Flannery, S. A. Teukolsky and W. T. Vetterling. 1986. *Numerical Recipes*. Cambridge University Press, p. 289.

Ogata, K. 1978. *System Dynamics*. Prentice Hall.

Tao, R., ed. 1991. *Electrorheological Fluids: Proceedings of the International Conference*, Carbondale.

Whittle, M. 1990. "Computer Simulation of an Electrorheological Fluid", *Journal of Non-Newtonian Fluid Mechanics*, 37:233–263.

Whittle, M., R. Firoozian, D. J. Peel and W. A. Bullough. 1991. "A Model for the Electrical Characteristics of an ER Valve", *Electrorheological Fluids: Proceedings of the International Conference*, Carbondale, pp. 343–366.

T - #0710 - 101024 - C0 - 276/216/8 - PB - 9781566761963 - Gloss Lamination